Advancements in Synovial Joint Science - Structure, Function, and Beyond

Edited by Alessandro Rozim Zorzi

Published in London, United Kingdom

Advancements in Synovial Joint Science - Structure, Function, and Beyond
http://dx.doi.org/10.5772/intechopen.1000412
Edited by Alessandro Rozim Zorzi

Contributors
Adnan Latif Malik, Ahlam Arnaout, Alessandro Rozim Zorzi, Camilo Chaves, Carlos Federico Morales Hackett, Efraín Navarro-Olivos, Enrico Carità, Fabio Andrés Tandioy-Delgado, Faiza Naseer, Faryal Jahan, Gilberto Flores-Vargas, Guadalupe Irazú Morales-Reyes, Imran Nazir, Ivan Tami, Jaime Sánchez Ruas, Jessica Paola Plascencia-Roldán, Kazunari Ninomiya, Keigo Nanjo, Lucas Alves Araújo, Lucian Lior Marcovici, Marco Guidi, María de Jesús Gallardo-Luna, Muhammad Imran Amirzada, Muhammad Soman Jahangir, Muhammad Umair Shahid, Nicolás Padilla-Raygoza, Rabia Gul, Safdar Iqbal, Saif ur Rehman, Satoshi Koto, Stefano Lucchina, Tahir Ahmad Baig, Takashi Ikeda, Wander Edney de Brito

First published in London, United Kingdom, 2024 by IntechOpen
IntechOpen is the global imprint of INTECHOPEN LIMITED, registered in England and Wales, registration number: 11086078, 5 Princes Gate Court, London, SW7 2QJ, United Kingdom

British Library Cataloguing-in-Publication Data
A catalogue record for this book is available from the British Library

Additional hard and PDF copies can be obtained from orders@intechopen.com

Advancements in Synovial Joint Science - Structure, Function, and Beyond
Edited by Alessandro Rozim Zorzi
p. cm.
Print ISBN 978-0-85466-479-5
Online ISBN 978-0-85466-478-8
eBook (PDF) ISBN 978-0-85466-480-1

For EU product safety concerns:
IN TECH d.o.o., Prolaz Marije Krucifikse Kozulić 3, 51000 Rijeka, Croatia,
info@intechopen.com or visit our website at intechopen.com.

Meet the editor

Prof. Alessandro R. Zorzi is a distinguished orthopedic surgeon with an extensive background in both clinical practice and academic research. He graduated in Medicine from the Faculdade de Medicina de Ribeirão Preto at the Universidade de São Paulo (USP), Brazil, a prestigious institution known for its rigorous training and excellence in medical education. He further advanced his expertise with an MSc and Ph.D. in Surgical Sciences from the University of Campinas (UNICAMP), Brazil. Additionally, he completed specialized training in clinical research at Harvard Medical School, USA, and in research ethics through UNESCO. Currently, Dr. Zorzi holds prominent positions at São Leopoldo Mandic, Brazil, where he is a dedicated researcher and professor in both undergraduate Medicine and Graduate Studies in Minimally Invasive Surgery and Translational Medicine. He also leads as the coordinator of the Center for Rare Diseases, underscoring his commitment to advancing medical knowledge and patient care in specialized fields. Within the broader medical community, Dr. Zorzi is highly regarded for his contributions. He practices as a physician in the Department of Orthopedics and Rheumatology at UNICAMP and in the Medical Assistance Group for Knee Surgery at Hospital Israelita Albert Einstein, highlighting his multifaceted expertise in orthopedics.

Contents

Preface

In *Advancements in Synovial Joint Science – Structure, Function, and Beyond*, we embark on an exploratory journey into the intricate world of synovial joints, aiming to shed light on the latest research, discoveries, and innovations that are shaping our understanding and treatment of these vital components of the human musculoskeletal system.

The genesis of this book was inspired by the recognition of the complex interplay between the structure and function of synovial joints and how this relationship underpins not only basic human movement but also the quality of life. Synovial joints, with their unique ability to provide mobility and bear the body's weight, are fundamental to our daily activities. However, they are also susceptible to a range of disorders and injuries that can severely impair functionality and wellness. It is this dichotomy that has fueled decades of scientific inquiry and clinical research, driving forward our collective knowledge and approaches to care.

This volume is designed for scientists, researchers, clinicians, and students who share a fascination with the biomechanics, pathology, and therapeutic strategies related to synovial joints. By compiling chapters that cover a wide spectrum of topics, we aim to provide a comprehensive overview that reflects the current state of the field while also looking toward future possibilities.

Each chapter of the book is penned by leading experts who bring not only their deep knowledge but also their perspectives on where their research may lead us in the coming years. This approach ensures that the content is not only grounded in the latest scientific evidence but also enriched with insights into potential breakthroughs and emerging challenges.

Advancements in Synovial Joint Science – Structure, Function, and Beyond is more than just a compilation of current knowledge; it is an invitation to engage with the ongoing quest for deeper understanding and better treatments. As you turn these pages, we hope you find not only information but also inspiration to contribute to this dynamic and ever-evolving field.

Alessandro Rozim Zorzi
São Leopoldo Mandic Medical School,
Campinas, Brazil

Chapter 1

Cartilage: Structure, Function, and the Pathogenesis of Osteoarthritis

Saif ur Rehman, Safdar Iqbal, Muhammad Umair Shahid, Muhammad Soman Jahangir and Adnan Latif Malik

Abstract

This chapter provides an in-depth exploration of cartilage, a pivotal component crucial for joint health, particularly within the context of osteoarthritis (OA). It delves deeply into the intricate structure and dynamic functions of articular cartilage, elucidating its essential roles in load-bearing, shock absorption, and maintaining joint stability. Emphasizing the delicate balance of cellular components, such as chondrocytes, and extracellular matrix constituents like proteoglycans and collagens, which collectively ensure the mechanical and biological integrity of cartilage, the discussion places significant attention on factors influencing cartilage homeostasis and contributing to its eventual degradation, analyzing age-related changes, mechanical stress, and genetic predispositions, alongside the impact of inflammatory processes and cytokine imbalances. By highlighting the multifaceted interplay among these factors, a clear narrative emerges, elucidating the initiation of OA. Furthermore, the chapter investigates into the cascade of events that define OA pathogenesis, dissecting the underlying mechanisms responsible for cartilage damage and matrix degradation, tracing their combined contribution toward the development of OA. In its focus on OA's etiology, the chapter underscores importance of comprehending cartilage alterations as a critical starting point for designing therapeutic interventions aimed at effectively managing OA.

Keywords: cartilage structure, cartilage function, cartilage degradation, cartilage repair, osteoarthritis pathogenesis

1. Introduction

Cartilage, a specialized type of connective tissue, plays a fundamental role in providing structural support and maintaining the integrity of various body structures. It exists in several forms, with each type adapted to specific functions and anatomical locations. The three main types of cartilage include hyaline cartilage, elastic cartilage, and fibrocartilage. Hyaline cartilage, recognized by its smooth and glassy appearance, occupies areas like the articular surfaces of joints. It functions as a protective cushion between bones, facilitating frictionless movements. Fibrocartilage, in contrast, boasts resilience and durability due to its high collagen fiber density. This type is prevalent in structures like intervertebral discs, contributing to shock absorption and stability. Elastic cartilage, containing both elastic fibers and collagen, is situated in regions requiring a blend of flexibility and support. Examples include the external ear and the epiglottis [1].

Articular cartilage, a specialized form of connective tissue, envelops the articulating surfaces of bones within synovial joints. It holds a pivotal role in enabling fluid and frictionless motion, guaranteeing joint stability, and effectively distributing mechanical loads across the joint surfaces. On one side, the articular cartilage interfaces with the joint cavity, providing the essential low-friction surface required for seamless joint movement. On the opposing side, it connects to the underlying subchondral bone plate *via* a delicate layer of calcified cartilaginous tissue, thereby upholding both structural integrity and stability [2, 3].

Referred to as hyaline cartilage due to its glassy and translucent appearance under microscopic observation, this specialized tissue boasts a distinctive and organized structure, which, when combined with its highly specialized composition, endows it with remarkable resilience against high compressive forces. These biophysical properties play a critical role in upholding the mechanical strength and durability of joints, particularly during weight-bearing activities. The amorphous nature of the articular cartilage, coupled with its lubricated surface and capacity to withstand substantial compressive forces, grants it the ability to withstand the repetitive movements and mechanical stresses inherent in joint motion. The absence of this smooth, wear-resistant cartilage surface would jeopardize joint functioning, leading to heightened friction and wear and the potential onset of painful joint conditions [3, 4].

Understanding articular cartilage is of paramount importance for unveiling the mechanisms that underlie joint diseases, notably osteoarthritis (OA), characterized by the gradual deterioration of cartilage. This condition impacts millions worldwide, serving as a major source of pain, disability, and compromised quality of life. Through the exploration of cellular and molecular processes governing cartilage health and degeneration, researchers can gain insights into the pathophysiology of OA, potentially paving the way for innovative therapeutic strategies. Moreover, advancements in regenerative medicine and tissue engineering hinge on a profound comprehension of articular cartilage. Crafting effective treatments for cartilage injuries and ailments demands a grasp of cartilage biology, biomechanics, and the intricate interplay between cells and the extracellular matrix components. Against the backdrop of mounting demand for joint-preserving interventions, the study of articular cartilage assumes pivotal significance, guiding the creation of interventions aimed at restoring its structure and function, thus mitigating the necessity for invasive joint replacement surgeries. Furthermore, as the global population ages and musculoskeletal disorders become more prevalent, a growing necessity emerges for early detection and intervention to prevent or delay cartilage degeneration. Research endeavors into noninvasive imaging techniques, biomarkers, and risk factors associated with cartilage deterioration hold the potential to identify individuals at risk of joint diseases, enabling timely interventions and personalized treatment strategies.

In conclusion, the study of articular cartilage transcends its structural and functional implications. It serves as the gateway to deciphering joint diseases, advancing regenerative therapies, and elevating the well-being of individuals dealing with joint-related conditions. By illuminating the intricate biology of this pivotal tissue, researchers and healthcare practitioners lay the groundwork for more efficacious treatments, early interventions, and improved joint health on a global scale.

2. Structure

The articular cartilage consists of chondrocytes, which are its resident cells, along with an extracellular matrix primarily comprised of collagen fibers, proteoglycans, and water. This combination, along with the pericellular matrix of chondrocytes, is collectively termed as the chondron [3]. Refer to **Figure 1** for an illustration of the structure of articular/hyaline cartilage.

2.1 Extracellular matrix (ECM)

Water dominates the extracellular matrix (ECM) of articular cartilage, constituting approximately 65–80% of its wet weight. As the cartilage traverses various zones, the water content decreases from roughly 80% in the superficial zone to around 65% in the deep zone [1, 2]. Water's pivotal role in upholding cartilage's structure and function cannot be overstated. A core function of water is hydrating the proteoglycans within the ECM. These proteoglycans, in partnership with water molecules, facilitate the expansion of the collagen network, endowing the cartilage with resilience and flexibility. Moreover, water functions as a lubricant within the joint, promoting smooth movement, while also aiding in the transportation of vital

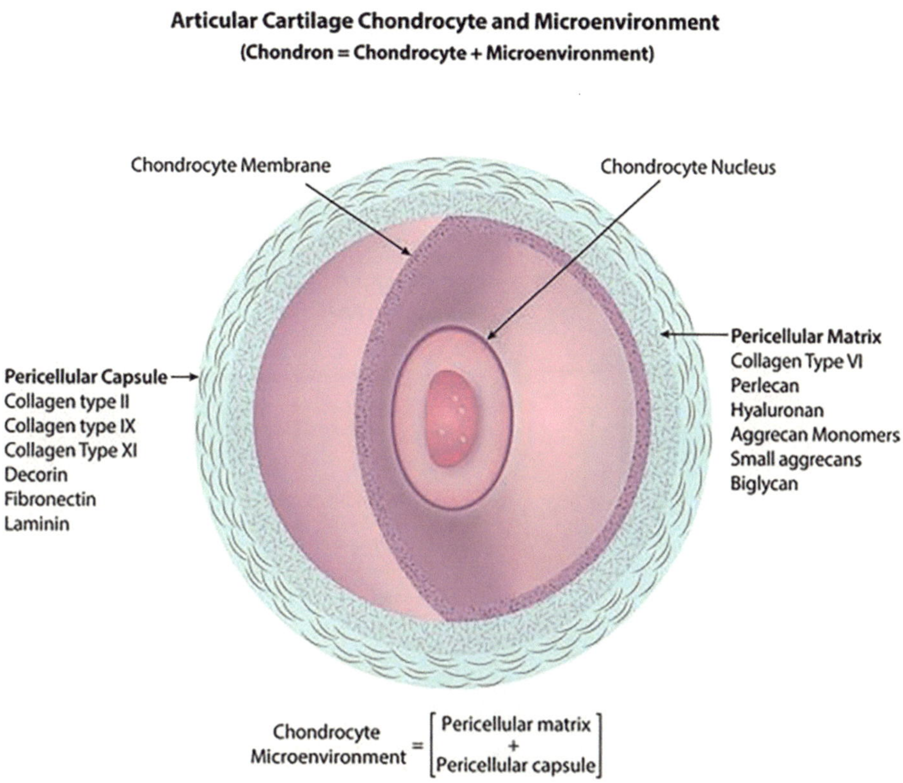

Figure 1.
A chondrocyte along with its surrounding microenvironment has been depicted in this picture. The membrane of the chondrocyte is encompassed by a delicate pericellular matrix consisting of type VI collagen, along with various minor glycoproteins and proteoglycans. This pericellular matrix is enveloped by a fibrous pericellular capsule composed of collagens and non-collagen proteins [2].

nutrients to cartilage cells [3]. The presence of water in the ECM is meticulously regulated through interactions with proteoglycans and collagens. Electrolytes, including sodium, potassium, chloride, and calcium, are integral components of this aqueous medium, further contributing to the maintenance of cartilage's biochemical equilibrium [1]. Over a person's lifespan, the water content in articular cartilage tends to diminish. However, individuals grappling with osteoarthritis (OA) may witness an increase in water content to approximately 90%. This surge in water content is linked to alterations in the mechanical properties of cartilage, resulting in reduced strength and heightened permeability (as depicted in **Figure 2**) [3, 4].

The second most abundant component of the articular cartilage's extracellular matrix is collagen. Chondrocytes, the specialized cells within the cartilage, secrete collagen in the form of procollagen molecules. These procollagen molecules undergo enzymatic cleavage of their C- and N-propeptides within the extracellular matrix (ECM). This processing step is essential for the normal growth of collagen fibrils. Intriguingly, even N-procollagen that has undergone partial processing can still come together to form slender collagen fibrils. Upon entering the extracellular matrix (ECM), these collagen molecules engage with one another and collaboratively polymerize to establish a fibrillar structure. The enduring structural integrity of these collagen fibrils is subsequently reinforced through the formation of covalent cross-links. These cross-links happen between adjoining collagen chains (referred to as intramolecular cross-links) and between neighboring collagen molecules (known as intermolecular cross-links). These cross-links provide a robust and resilient structure, crucial for maintaining the integrity and mechanical properties of the articular cartilage [3]. In human articular cartilage, collagen type II is the primary structural protein, making up approximately 90–95% of the total collagen content. The remaining 5–10% of collagen is composed of other cartilage-specific and nonspecific collagens, accounting for about 1% of the cartilage's dry weight (**Figure 3**) [5].

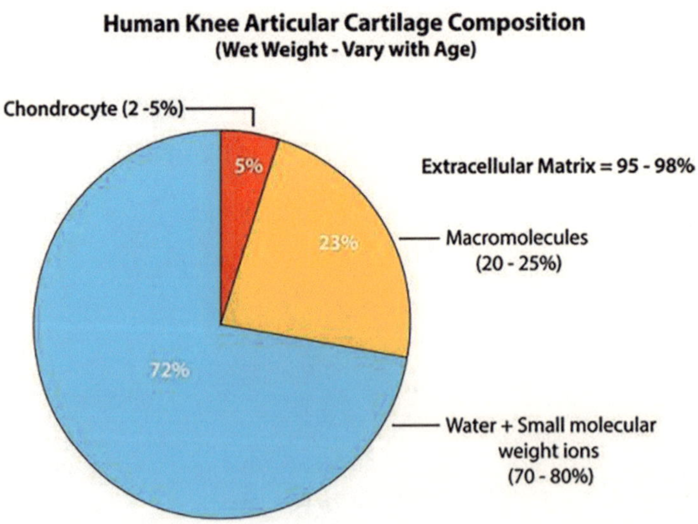

Figure 2.
The composition of wet weight in the knee joint cartilage of the human adult reveals that chondrocytes constitute a mere 2–5% of the total weight, while the bulk of 95–98% is composed of the extracellular matrix. This matrix is largely comprised of tissue fluid and low molecular weight ions. It is worth noting that the wet weight of the macromolecules in knee articular cartilage differs based on both age and the specific anatomical location [2].

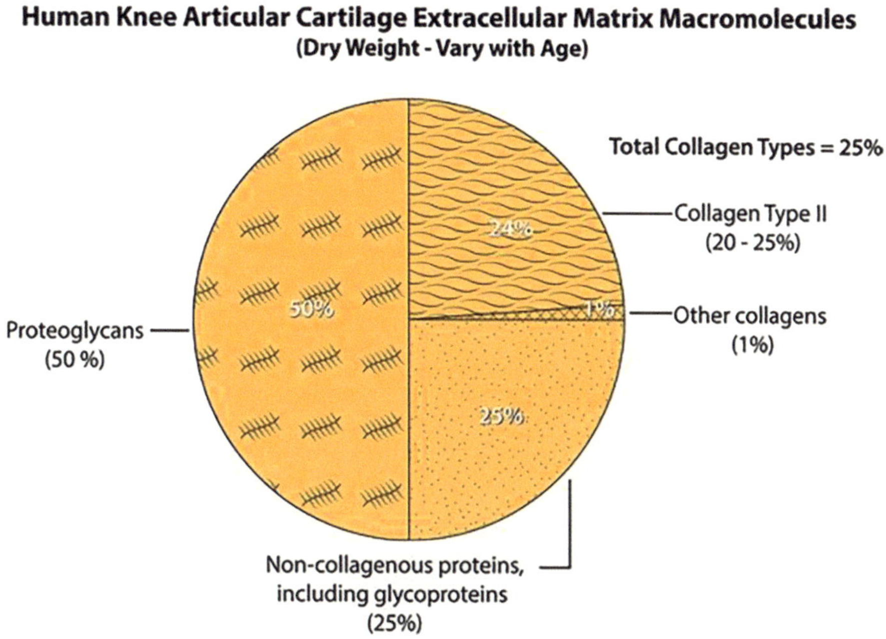

Figure 3.
The composition of macromolecules by dry weight in the extracellular matrix of hyaline cartilage within the human adult knee. It is important to observe that the dry weight of these molecules within the articular cartilage fluctuates based on both age and the specific anatomical location [2].

Interestingly, the turnover rate of collagen type II in normal adult cartilage is exceedingly slow, with a half-life of over 100 years. In contrast, proteoglycans (PGs) and aggrecans, which are essential components of the cartilage's extracellular matrix (ECM), continue to be produced and secreted throughout life. The usual turnover rate for a large monomer of PG corresponds to a half-life of about 3.4 years [6]. This dynamic balance between collagen and proteoglycan turnover is crucial for maintaining cartilage structure and function. Besides collagen and proteoglycans, other non-collagenous proteins play a significant role in the ECM's composition, constituting approximately 25% of the dry weight depending on age. These proteins include laminin, fibronectin, chondronectin, tenascin, cartilage matrix glycoprotein (CMGP), and cartilage oligomeric matrix protein (COMP). These glycoproteins contribute to the structural organization and integrity of the ECM, promoting cell–matrix interactions and tissue stability [1]. Collagen derives its impressive tensile strength from a unique triple-helix structure, where three polypeptide chains are intertwined through hydrogen bonds. Each polypeptide consists of an iterating sequence of amino acids: proline, glycine, and hydroxyproline, forming a left-handed helical structure through additional hydrogen bonds [7]. Although Type II collagen is the most studied collagen, yet other collagen types, such as Types IV, VI, IX, X, XI, XII, XIII, and XIV, also exist in articular cartilage. These less-studied collagen fibers might offer valuable insights into disease progression and potential biomarkers for improved treatment [3–5]. Collagen is distributed throughout the extracellular matrix, and its arrangement varies depending on the regional differences within the joint cartilage (different zones). Additionally, the density of collagen is higher at the apical surface of a chondron compared to at the basal side [5]. Collagen forms associations and cross-links

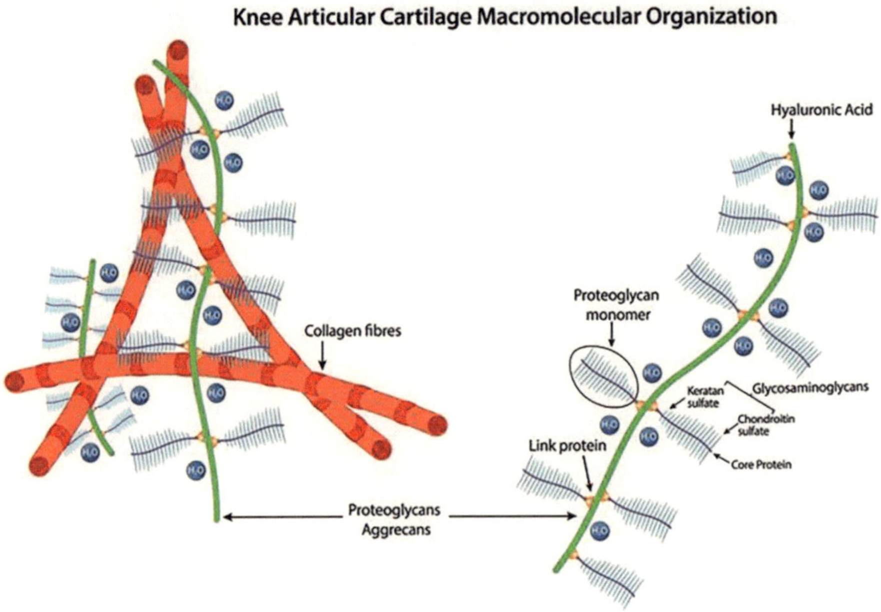

Figure 4.
The picture depicts interaction and arrangement of collagen and aggrecan within the articular cartilage. The basic structure of the proteoglycan monomer comprises a central protein core that is linked through covalent bonds to glycosaminoglycan side chains, specifically keratan sulfate and chondroitin sulfate. These monomers are connected to the hyaluronic acid back through a region that binds hyaluronic acid (referred to as the hyaluronic acid-binding region), which is additionally stabilized by the presence of the link protein [2].

with proteoglycans, creating the fundamental organizational unit of the extracellular matrix (**Figure 4**) [3].

Proteoglycans are another essential component of the extracellular matrix and are found throughout the connective tissues. They contribute to the strength of the matrix and are hydrophilic due to their negative charges. This hydrophilic property plays a crucial role in facilitating the lubrication of the joint's bearing surfaces, ensuring smooth and frictionless movement during joint motion, minimizing wear and tear on the cartilage surface. Proteoglycans are proteins covalently linked to glycosaminoglycans (GAGs), which are repetitive dimers of a hexosamine and a uronic acid [3, 7]. In articular cartilage, proteoglycans exhibit significant variability in their size, glycosaminoglycan (GAG) content, and functional properties. The proteoglycans (PGs) can be classified into two major groups based on their properties and functions. The first group comprises large aggregating PGs, which include molecules like aggrecan and versican. These large aggregating PGs are responsible for forming complex structures that can trap and retain a significant amount of water, backing to the cartilage's ability to endure compressive forces and maintain its elasticity. They make up a substantial portion of the total PGs, accounting for about 50–58%. The second group consists of non-aggregating PGs, making up approximately 40% of the total PGs [7, 8]. Unlike the large aggregating PGs, non-aggregating PGs do not form extensive complexes with water. Instead, they serve other crucial functions within the cartilage matrix. These non-aggregating PGs play roles in cell signaling, tissue organization, and interacting with other components of the extracellular matrix. The balance between these two types of PGs is essential for maintaining the overall structure and function of articular cartilage [3, 4, 8]. In the case of aggrecan, the major GAGs attached are chondroitin sulfate and keratin sulfate. Hyaluronic acid (HA) is another crucial GAG for the function and structure of articular cartilage. Although it does not form covalent bonds to proteins like other proteoglycans, yet hyaluronic acid plays a significant role by making non-covalent complexes with proteoglycans, particularly aggrecan, through proteoglycan link proteins. Collectively, hyaluronic acid and proteoglycans form extensive proteoglycan-hyaluronic acid aggregates that fix to the surface of collagen II fibers by their side chains, creating strong connections among all ECM constituents and forming the resilient backbone of hyaline cartilage [8].

2.2 Chondrocytes

Chondrocytes represent the vital cellular component of cartilage that is housed within small spaces called lacunae. Despite their significance, chondrocytes account for only about 5% of the total volume of articular cartilage [9]. These chondrocytes exhibit diverse morphologies, transitioning from flat, discoid-shaped cells at the surface of the cartilage to round or polygonal shapes as we move deeper into the cartilage tissue. These spheroidal cells are not uniformly distributed but rather form clusters known as isogenous groups, and their metabolic activity plays a vital role in sustaining the integrity of the extracellular matrix surrounding them [9]. One of the unique aspects of chondrocytes' environment is its hypoxic nature, meaning it has a low oxygen concentration. As a result, a considerable portion of the chondrocytes' metabolic processes is anaerobic, not reliant on oxygen [3, 4]. These anaerobic pathways are adapted to the specific conditions of the cartilage, allowing chondrocytes to thrive in this oxygen-deprived environment [3]. Chondrocytes originate from mesenchymal stem cells, which differentiate into specialized cells called chondroblasts. The chondroblasts then play a critical role in the formation and secretion of the essential components of the ECM, such as collagens and proteoglycans [10]. These ECM components provide the cartilage with its structural framework and mechanical properties. As chondroblasts continue to synthesize the ECM, they eventually become surrounded and completely embedded by the matrix they produce. At this point, they are referred to as chondrocytes [8, 10]. Once chondrocytes are fully enclosed within their lacunae, they remain metabolically active and continue to maintain the cartilage's health and function. Overall, chondrocytes are vital cellular units within cartilage responsible for ECM maintenance and ensuring proper cartilage structure and function. Their ability to adapt to the unique hypoxic environment of cartilage and the ongoing metabolic processes is crucial for the long-term health and integrity of articular cartilage [3, 4]. Understanding the behavior of chondrocytes and their role in cartilage physiology is essential for developing effective therapies to address cartilage-related disorders and promote joint health.

2.3 Zones

Articular cartilage can be divided into distinct zones based on their unique structural and functional characteristics (**Figure 5**) [11].

Superficial (*tangential*) *zone*: The superficial zone is the thin outermost layer of articular cartilage, comprising approximately 10–20% of its total thickness [11–13]. This zone plays a crucial role in protecting the deeper layers from shear stresses. The collagen fibers in this zone, mainly type II and IX collagen, are densely packed and oriented parallel to the articular surface. Flattened chondrocytes are relatively abundant in this layer. The integrity of the superficial zone is vital for safeguarding

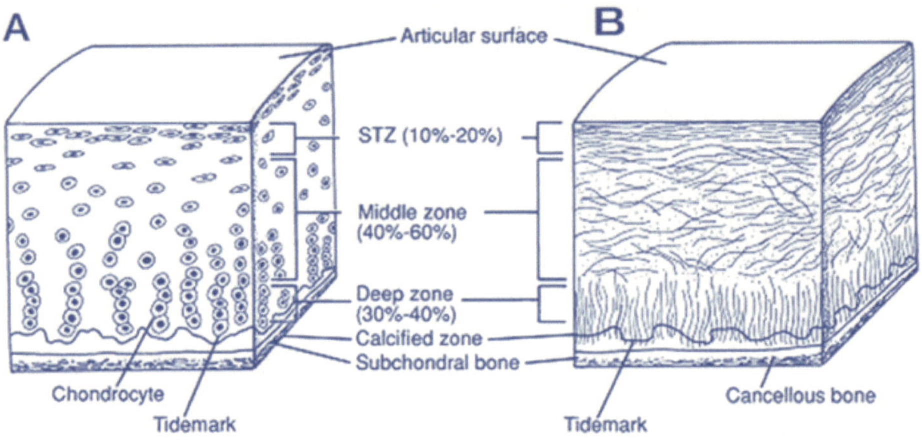

Figure 5.
A cross-sectional schematic diagram of intact articular cartilage is displayed, illustrating two main aspects: (A) the cellular arrangement across different zones within the articular cartilage structure and (B) the intricate architecture of collagen fibers [3].

and maintaining the health of the underlying cartilage layers. It is also responsible for providing most of the tensile properties of cartilage, enabling it to resist the tensile, shear, and compressive forces experienced during joint movement [11, 12].

Middle (transitional) zone: The middle zone lies immediately beneath the superficial zone and serves as a functional bridge between the superficial and deep zones. It constitutes approximately 40–60% of the total cartilage volume [13]. In this layer, the collagen fibers are organized obliquely, and the chondrocytes are spherical and less densely distributed compared to the superficial zone. The middle zone is primarily responsible for resisting compressive forces. It contains proteoglycans and thicker collagen fibrils, providing support and mechanical resistance during joint loading [11–13].

Deep zone: The deep zone is the deepest layer of articular cartilage and accounts for around 30% of its volume [13]. This zone plays a critical role in providing the highest resistance to compressive forces. The collagen fibrils in the deep zone are arranged perpendicular to the articular surface, conferring optimal strength for withstanding compressive loads. The deep zone contains the largest diameter collagen fibrils, the highest proteoglycan content, and the lowest water concentration. Chondrocytes in this layer are typically arranged in a columnar orientation, parallel to the collagen fibers and perpendicular to the joint line. The deep zone is essential for maintaining the structural integrity and mechanical resilience of articular cartilage during weight-bearing activities [11, 12].

Tide mark and calcified cartilage: The tide mark defines the boundary between the deep zone and the calcified cartilage, which is a transition zone between the cartilage and subchondral bone. The calcified cartilage plays a vital role in anchoring the collagen fibrils of the deep zone to the subchondral bone, providing a stable connection between cartilage and bone. Chondrocytes in this zone are scarce and often hypertrophic [11, 12].

To conclude, the zonal organization of articular cartilage enables it to withstand the various mechanical stresses imposed during joint movement. Each zone possesses distinct structural characteristics and functions, contributing to the overall mechanical properties and resilience of articular cartilage in maintaining smooth joint articulation and joint health [3, 13].

2.4 Regions

The extracellular matrix (ECM) of articular cartilage exhibits distinct regions based on their proximity to chondrocytes, composition, and organization of collagen fibrils. These regions can be categorized into pericellular, territorial, and interterritorial regions [3, 4].

The *pericellular matrix* is a thin layer surrounding the cell membrane of chondrocytes. It completely encases the chondrocyte and is primarily composed of proteoglycans, glycoproteins, and other non-collagenous proteins. This matrix region is believed to play a functional role in initiating signal transduction within the cartilage in response to mechanical loading. It may serve as a key communicator between the chondrocytes and their surrounding environment, influencing cellular activities and tissue homeostasis [3].

The *territorial matrix* surrounds the pericellular matrix and is mostly made up of fine collagen fibrils, creating a basket-like network around the chondrocytes. This area is much thicker than the pericellular matrix and is thought to offer protection to the cartilage cells against mechanical stresses. Additionally, the territorial matrix may contribute to the overall resiliency of the articular cartilage structure, enabling it to withstand substantial mechanical loads and forces during joint movement [3].

The *interterritorial region* is the largest of the three matrix regions and plays a significant role in the biomechanical properties of articular cartilage. It contributes extensively to the cartilage's mechanical strength and durability. In this region, large collagen fibrils are arranged in randomly oriented bundles. Their orientation varies in different zones of articular cartilage, being parallel to the surface in the superficial zone, oblique in the middle zone, and perpendicular to the joint surface in the deep zone. Abundant proteoglycans are present in the interterritorial zone, contributing to its ability to resist compressive forces and provide structural support during weight-bearing activities [3].

In summary, the ECM of articular cartilage can be subdivided into pericellular, territorial, and interterritorial regions, each with unique compositions and functions. These distinct regions work in harmony to maintain the cartilage's mechanical integrity, ensuring smooth joint movement and providing essential protection and support to chondrocytes in response to various mechanical stresses.

3. Functions of articular cartilage

Cartilage plays a paramount role in joint function and mobility; operating as a silent yet essential partner, it orchestrates an array of indispensable mechanisms that are pivotal for sustaining seamless and pain-free movement. It operates as a cushion, a shock absorber, a load distributor, and a friction-reducing agent, all of which contribute to the smooth functioning of joint.

3.1 Smooth articulation

One of the primary functions of articular cartilage is to facilitate smooth articulation between bones in the joints. By preventing bone-to-bone contact, cartilage ensures that the joint surfaces can move smoothly against each other. During joint movement, such as flexion and extension, the smooth articulation provided by cartilage reduces the impact and shear forces on the joint. This helps to prevent excessive wear and tear on the cartilage and underlying bones. Without the protective cushioning and smoothness of cartilage, joint surfaces would be subjected to direct contact, leading to damage, inflammation, and pain over time [14].

3.2 Load distribution

Articular cartilage serves as a load-bearing surface within the joints, by distributing the mechanical forces and loads that occur during movement. This function is essential for preventing excessive stress on specific areas of the joint and ensuring uniform force distribution across the entire joint surface. During weight-bearing activities like walking, running, or lifting, the joints experience substantial compressive forces. These forces can be quite significant, especially in weight-bearing joints like the knees and hips. Without the load distribution function of articular cartilage, the joint surfaces would be subject to concentrated pressure, which could lead to local damage, wear, and ultimately joint degeneration. The collagen fiber network within the cartilage provides structural support and enhances its load distribution capacity. Collagen fibers are arranged in a specific pattern, enabling the cartilage to resist tensile and shear forces effectively. These fibers work in conjunction with the proteoglycans to disperse loads evenly across the joint surface [7, 14].

3.3 Low-friction surface

The unique composition of cartilage, particularly the presence of proteoglycans and synovial fluid, creates a low-friction surface within the joint. Cartilage is characterized by a very low coefficient of friction, both static (0.01–0.02) and dynamic (0.003). This minimizes the resistance during joint movement, allowing for smooth and effortless motion. As a result, cartilage enables a wide range of movements, including flexion, extension, rotation, and gliding, which are essential for activities of daily living and physical performance. This lubrication minimizes wear and tear on the joint surfaces, promoting longevity and preserving joint function [14].

3.4 Shock absorption

Cartilage's ability to absorb shock is vital for protecting the joints and preventing injury during weight-bearing activities and impact-related movements. Articular cartilage may be collapsed by up to 40% of its resting height when subjected to physiological stresses [4]. This is secondary to its high elasticity and compressibility in turn provided by the collagen fiber network as well as proteoglycans owing to their capacity to bind water molecules and cations, allowing them to maintain high osmotic pressure within the cartilage [3, 4]. When pressure is applied to the joint during activities such as jumping or lifting heavy objects, the water within the cartilage matrix is forced out. This redistribution of fluid helps counteract the compressive forces and maintain joint stability. As the pressure is released, the water is reabsorbed, allowing the cartilage to return to its original shape and ensuring proper joint function. By absorbing and dissipating forces, cartilage reduces the risk of damage to both the joint surfaces and the surrounding structures. This shock-absorbing property is especially important in high-impact activities, such as running or jumping, where joints are subjected to increased stress [7, 14, 15].

3.5 Joint stability

Healthy cartilage plays a vital role in maintaining joint stability and overall joint health. Acting as a natural cushion between bones, cartilage prevents direct contact between joint surfaces during movement. This cushioning effect is crucial for reducing friction and wear on the bones, safeguarding them from damage and ensuring their longevity. Moreover, the smooth surface of cartilage allows bones to articulate seamlessly, further enhancing joint stability and preventing dislocations or subluxations [14].

In summary, cartilage is an indispensable tissue in joint function and mobility. Its ability to facilitate smooth articulation, distribute loads, provide a low-friction surface, absorb shock, and maintain joint stability is crucial for overall musculoskeletal health. Proper care and preservation of cartilage are essential to sustain joint function and prevent the onset of joint disorders, such as osteoarthritis, which can significantly impact an individual's quality of life.

4. Factors influencing cartilage repair and regeneration

Cartilage has limited regenerative capacity due to its avascular nature and low cellular density. Nevertheless, cartilage repair and regeneration can occur to some extent under specific conditions. Several intrinsic and extrinsic factors influence the repair and regeneration processes. Understanding these factors is essential for developing strategies to enhance cartilage healing and to potentially address cartilage-related disorders such as osteoarthritis.

4.1 Age

Cartilage is known for its limited ability to self-repair, and this regenerative capacity diminishes with advancing age. Younger individuals generally exhibit better cartilage-healing capabilities compared to older individuals. Several factors contribute to this age-related decline in cartilage repair. One of the main reasons for reduced cartilage repair in older individuals is the decline in the metabolic activity of chondrocytes, the specialized cells responsible for maintaining cartilage. Chondrocytes play a vital role in synthesizing and maintaining the extracellular matrix (ECM) of cartilage, which includes collagen fibers and proteoglycans. As age progresses, the chondrocytes become less active and have a slower rate of ECM synthesis, thereby impairing their ability to repair damaged cartilage. Clinical evidence supports the notion that cartilage-healing capabilities decline with age. For example, studies have shown that cartilage injuries in young adults tend to heal more efficiently and produce better-quality repair tissue than in older adults [16]. A study published in the American Journal of Sports Medicine assessed cartilage repair outcomes in patients undergoing autologous chondrocyte implantation (ACI) for knee cartilage defects. The researchers found that younger patients (under 40 years old) achieved better outcomes in terms of cartilage repair and clinical improvement compared to older patients (over 40 years old) [17]. As a result of the reduced regenerative capacity, cartilage injuries in older individuals are less likely to heal completely. Instead, the body may initiate a repair response that leads to the formation of fibrocartilage, which is a structurally inferior tissue compared to native hyaline cartilage [16]. Fibrocartilage lacks the organization and mechanical properties of hyaline cartilage, making it less effective in withstanding mechanical stresses and maintaining joint function [18].

The age-related decline in cartilage repair and regeneration has significant implications for joint health. As age-related factors contribute to reduced cartilage repair and regeneration, older individuals may be at a higher risk of developing osteoarthritis or experiencing more severe joint degeneration [19].

4.2 Severity of injury

The ability of cartilage to repair itself is limited due to its avascular nature [3, 4, 20]. The extent and severity of cartilage injury indeed play a crucial role in the repair process. Minor injuries, such as small defects or superficial lesions, generally have a better chance of healing compared to more extensive injuries. For instance, a study published in the Journal of Orthopaedic Research examined the repair of small, full-thickness cartilage defects in the knee joint in rabbits. The researchers found that these small defects exhibited better healing outcomes compared to larger defects, with evidence of new cartilage formation and tissue integration [21]. In contrast, larger and deep cartilage injuries pose greater challenges to the repair process. Injuries that extend into the deeper layers of cartilage may involve damage to the underlying subchondral bone or even reach the bone itself. The lack of direct blood supply to cartilage and the limited ability of cartilage cells (chondrocytes) to proliferate and migrate to the injury site hinder the repair process. Consequently, large and deep injuries are less likely to heal spontaneously and may result in incomplete or inadequate cartilage repair. The presence of a stable and intact subchondral bone is also crucial for cartilage repair as it provides a scaffold for cell migration and tissue repair. It serves as a source of growth factors and other signaling molecules that can stimulate cartilage repair processes [20]. Clinical studies have emphasized the importance of the subchondral bone in cartilage repair. The researchers have found that when subchondral bone integrity was compromised, cartilage repair was less successful, and fibrocartilage was formed instead of hyaline cartilage [22].

4.3 Blood supply

Cartilage's avascular nature poses a significant challenge to its healing process. Unlike well-vascularized tissues, cartilage relies on alternative mechanisms for repair, limiting its regenerative capacity. Without blood vessels, essential nutrients, oxygen, and immune cells cannot efficiently reach the injured site, hampering chondrocytes' access to resources needed for effective healing [3, 4]. The lack of immune cells at the injury site also reduces the body's ability to initiate the inflammatory and reparative responses necessary for tissue healing [5]. However, there are certain scenarios in which cartilage injuries may extend into the subchondral bone. The subchondral bone is well-vascularized, containing numerous blood vessels that supply nutrients and oxygen to the underlying bone and surrounding tissues. In some cases, when an injury extends into the subchondral bone, blood vessels from the bone may penetrate the injured cartilage, potentially improving the healing response [21, 22]. Clinical studies have shown that cartilage injuries that extend into the subchondral bone may exhibit improved healing potential compared to injuries limited to the cartilage layer alone [22].

4.4 Mechanical stimulation

Mechanical loading plays a crucial role in the repair and regeneration of cartilage. Cartilage is a dynamic tissue that responds to mechanical forces by adapting its structure and function. Proper mechanical stimulation through controlled joint movement and physical activity is essential for cartilage health and maintenance. Controlled and appropriate joint movement helps distribute nutrients and fluid within the cartilage, providing essential nourishment to the chondrocytes and maintaining the extracellular matrix (ECM) integrity. This mild mechanical loading stimulates chondrocyte activity and ECM turnover, promoting tissue repair and remodeling [23, 24]. In a study published in the American Journal of Sports Medicine, researchers investigated the impact of controlled dynamic compression on cartilage repair. They found that mild mechanical loading led to increased chondrocyte proliferation and synthesis of ECM components, suggesting a potential role in cartilage repair and regeneration [25]. On the other hand, excessive or abnormal mechanical loading can be detrimental to cartilage health. Overloading the joint with excessive forces can lead to cartilage degeneration and damage. High-impact activities or repetitive loading beyond the physiological capacity of the joint can cause wear and tear on the cartilage, leading to the breakdown of ECM components, thereby compromising cartilage integrity [26].

4.5 Inflammation

Inflammatory processes indeed play a critical role in cartilage repair and regeneration. Inflammation is a natural and necessary response of the body to tissue injury, including cartilage damage. It is an essential early stage in the healing process, as it helps to initiate tissue repair and recruit immune cells and growth factors to the injured site [15]. When cartilage is damaged, due to either trauma, wear and tear, or degenerative conditions, the body's immune system responds by triggering an acute inflammatory response. This acute inflammation helps remove damaged tissue and debris, promotes the release of growth factors and cytokines, and attracts immune cells, such as macrophages and neutrophils, to the injury site [15]. Clinical evidence supports the role of acute inflammation in cartilage repair [27]. While acute inflammation is a beneficial and necessary step in cartilage repair, chronic inflammation can have detrimental effects on the healing process and cartilage health. In chronic inflammatory conditions, such as rheumatoid arthritis, release of destructive enzymes that break down the ECM components of cartilage disrupts the delicate balance between tissue breakdown and tissue repair [28] and contributes to progressive cartilage loss and joint destruction [29].

4.6 Growth factors and cytokines

Cartilage repair and regeneration are highly influenced by various growth factors and cytokines that regulate the behavior of chondrocytes. These signaling molecules play a critical role in orchestrating the complex processes involved in cartilage healing and tissue remodeling [15]. Transforming growth factor-beta (TGF-β) and insulin-like growth factor-1 (IGF-1) are two essential growth factors that stimulate chondrocyte activity and contribute to cartilage repair and regeneration. TGF-β is known for its potent effects on chondrocyte proliferation and differentiation. It promotes the synthesis of extracellular matrix (ECM) components, including collagen and proteoglycans, which are essential for cartilage structure and function. TGF-β also helps to regulate the balance between cartilage breakdown and synthesis, favoring the repair and rebuilding of damaged cartilage [30]. Insulin-like growth factor-1 (IGF-1) is another growth factor that plays a significant role in cartilage repair. It acts as a potent mitogen, stimulating chondrocyte proliferation and promoting the synthesis of ECM proteins. IGF-1 also enhances the anabolic activity of chondrocytes, leading to increased production of cartilage-specific molecules that are crucial for cartilage repair and regeneration [31]. In contrast to growth factors that promote cartilage repair, certain cytokines can have detrimental effects on cartilage health. Pro-inflammatory cytokines, such as interleukin-1 (IL-1), are key mediators of the inflammatory response and are often elevated in conditions like osteoarthritis and rheumatoid arthritis. IL-1 can promote cartilage degradation by stimulating the production of matrix-degrading enzymes, such as matrix metalloproteinases (MMPs) and aggrecanases. These enzymes can break down the ECM components of cartilage, leading to cartilage degradation and tissue damage. Additionally, IL-1 can inhibit the synthesis of ECM proteins, further compromising cartilage repair [32]. Clinical evidence has confirmed the impact of these growth factors in cartilage repair and degeneration [33].

4.7 Treatment interventions

The choice of treatment interventions can significantly impact cartilage repair and regeneration. Surgical techniques, such as microfracture, autologous chondrocyte implantation (ACI), or matrix-assisted autologous chondrocyte transplantation (MACT), aim to promote cartilage healing by stimulating the formation of fibrocartilage or transplanting healthy chondrocytes into the injured site [34]. Moreover, it is noteworthy that several treatments intended to enhance the quality of life for patients with mild to moderate arthritis might inadvertently contribute to cartilage damage. Notably, interventions such as intra-articular steroid and lignocaine injection, commonly administered to alleviate pain in individuals with arthritic knees, have been found to accelerate the deterioration of cartilage and the advancement of arthritis [35, 36]. Although these treatments may offer short-term pain relief and improved mobility, yet they also carry the risk of exacerbating cartilage breakdown over the long term, potentially impacting joint health and function.

4.8 Adjacent tissues and joint alignment

The health and stability of adjacent tissues, such as ligaments and menisci, play a crucial role in cartilage repair and overall joint health. These structures are intimately connected and work together to ensure proper joint function and stability. When adjacent tissues like ligaments and menisci are damaged or compromised, they can lead to altered joint biomechanics, causing abnormal loading and stress on the cartilage. This increased stress can contribute to cartilage degeneration and hinder the healing process of existing cartilage injuries. For example, a torn anterior cruciate ligament (ACL) in the knee can destabilize the joint, leading to abnormal movement and increased shear forces on the cartilage surfaces. This can result in cartilage damage and accelerate the progression of osteoarthritis. In contrast, a well-functioning ACL helps in maintaining joint stability and promotes more uniform distribution of forces, reducing the risk of cartilage injuries and degeneration [37, 38]. To support cartilage repair and regeneration, it is essential to address any concurrent injuries or instability in the joint. Restoring the health and stability of ligaments and menisci through appropriate rehabilitation or surgical intervention can optimize joint biomechanics and reduce excessive loading on specific areas of the cartilage. This, in turn, creates a more favorable environment for cartilage healing and helps to preserve joint function and longevity [39].

In conclusion, cartilage repair and regeneration are complex processes influenced by multiple factors. Age, severity of injury, blood supply, mechanical stimulation, inflammation, growth factors, treatment interventions, and joint alignment all play significant roles in determining the success of cartilage healing. Understanding and optimizing these factors is crucial for developing effective strategies to enhance cartilage repair, promote tissue regeneration, and potentially alleviate the impact of cartilage-related disorders.

5. Pathogenesis of osteoarthritis

Osteoarthritis (OA) is a complex and multifactorial degenerative joint disorder that affects millions of people worldwide. Articular cartilage is a crucial component of synovial joints, facilitating smooth and painless movement while providing load-bearing support. In osteoarthritis (OA), the progressive degeneration of articular cartilage is a central feature of the disease's pathogenesis. The breakdown of this essential tissue leads to joint pain, stiffness, and functional impairment. This section explores the intricate processes involved in articular cartilage degeneration in OA, shedding light on the key factors contributing to its deterioration.

5.1 Role of mechanical stress

Mechanical stress stands as a pivotal driving force in the intricate interplay of factors contributing to articular cartilage degeneration, particularly in the context of osteoarthritis (OA). The complex nature of joint movement orchestrates a symphony of compressive and shear forces that cartilage experiences during daily activities and weight-bearing tasks. In a healthy joint, the remarkable resilience of cartilage enables it to adeptly manage and distribute these mechanical forces, maintaining the joint's structural integrity and function [40]. However, in the complicated tapestry of osteoarthritis, this equilibrium is disrupted. The delicate balance between mechanical stress and the cartilage's inherent capacity to regenerate and adapt becomes compromised. Excessive or aberrant mechanical loading, commonly observed in OA due to altered joint mechanics, musculoskeletal imbalances, or other contributing factors, manifests as a critical catalyst in cartilage degeneration [40, 41]. The persistent or abrupt mechanical stress leads to microdamage within the cartilage structure, creating fissures, cracks, and areas of localized stress concentration. This microdamage is a tipping point that prompts a complex series of biochemical responses within the cartilage tissue. In response to microdamage, chondrocytes initiate a complex web of signaling pathways. This includes the release of inflammatory mediators, matrix-degrading enzymes, and pro-inflammatory cytokines. The inflammatory milieu sets the stage for a heightened state of catabolism within the cartilage, promoting the degradation of matrix components like collagen and proteoglycans [41]. Over time, the sustained and cumulative impact of this cascade of events erodes the cartilage's structural integrity. As the cartilage matrix loses its components and the chondrocytes' regenerative capacity diminishes, the once-resilient tissue succumbs to progressive degradation and thinning. This, in turn, amplifies joint pain, restricts mobility, and exacerbates the degenerative process characteristic of osteoarthritis [42].

5.2 Proteoglycan loss

One of the early changes observed in OA is the loss of proteoglycans from the cartilage matrix. As discussed earlier, proteoglycans are essential for retaining water within the cartilage, which gives it its shock-absorbing properties. In the early stages of OA, the delicate equilibrium within the cartilage matrix becomes disrupted. The loss of proteoglycans is observed as a result of altered metabolism and biochemical changes in the cartilage tissue. This depletion of proteoglycans directly impacts the cartilage's capacity to retain water, leading to a reduction in its shock-absorbing capabilities. As the disease progresses, the diminished water retention and subsequent loss of proteoglycans compromise the cartilage's ability to withstand the mechanical stresses encountered during joint motion [19]. The consequence of this proteoglycan loss is twofold. First, the cartilage becomes less efficient at absorbing and distributing mechanical forces generated by activities such as walking, running, or weight-bearing. This diminished shock-absorbing capacity results in increased mechanical stress being transmitted directly to the underlying bone and joint tissues. Secondly, the altered biomechanical properties of the cartilage contribute to an environment conducive to further damage. The compromised cartilage becomes more susceptible to microdamage and fibrillation, which, in turn, accelerates the progression of OA [19, 43].

5.3 Collagen disorganization

Collagen fibers play a fundamental role in maintaining the robust structural integrity of articular cartilage. In healthy conditions, the ordered and aligned arrangement of collagen fibers within the cartilage matrix imparts resilience and durability, contributing significantly to the cartilage's ability to function as a cushioning buffer within the joint. However, the intricate balance of collagen organization becomes disturbed in the context of osteoarthritis (OA). This degenerative joint disorder introduces a series of detrimental changes that compromise the architecture of collagen fibers. This process is referred to as collagen disorganization, and it manifests as alterations in the arrangement, alignment, and density of collagen fibers within the cartilage matrix. The consequences of collagen disorganization in OA are far-reaching [44]. As collagen fibers lose their well-ordered configuration, the tensile strength that they once provided diminishes. The once-efficient mechanical network that facilitated even distribution of forces across the cartilage surface becomes disrupted. This weakening effect leaves the cartilage more vulnerable to the mechanical stress generated during joint movement [44, 45].

Consequently, areas of increased stress concentration form on the cartilage surface, which can lead to localized microdamage and tissue degradation. Furthermore, the altered alignment of collagen fibers results in reduced ability to resist shear forces. Shear forces, which occur when opposing surfaces slide against each other, place additional stress on the cartilage. The diminished capacity of disorganized collagen to withstand these forces exacerbates the wear and tear experienced by the cartilage that hastens its deterioration. The interaction between collagen disorganization and cartilage breakdown is a vicious cycle. As the cartilage matrix loses its integrity due to collagen disorganization, chondrocytes respond with an elevated production of matrix-degrading enzymes and inflammatory mediators [45, 46]. These biochemical changes contribute to the degradation of collagen and other components within the cartilage, further accelerating tissue damage and exacerbating the OA process [15].

5.4 Chondrocyte dysfunction

Chondrocytes, assume a pivotal role in upholding the overall health and integrity of the cartilage matrix. However, in the context of osteoarthritis (OA), the functioning of chondrocytes undergoes significant alterations, which contribute to the degenerative processes associated with this condition. In OA, chondrocytes experience a shift in their synthetic activity, marked by a decline in their ability to produce essential extracellular matrix components. This diminished synthetic capacity impairs the cartilage's ability to replenish its structural elements, thereby weakening its resilience and ability to withstand mechanical stresses. Moreover, in OA-affected joints, chondrocytes exhibit an augmented production of catabolic enzymes, particularly matrix metalloproteinases (MMPs) and aggrecanases. These enzymes act as molecular scissors, cleaving various components of the extracellular matrix, including collagen and proteoglycans. Such enzymatic breakdown of the matrix accelerates the degradation of cartilage tissue disrupting of the delicate balance between matrix synthesis and degradation within the cartilage. As the synthetic capacity wanes and catabolic enzyme activity increases, the extracellular matrix deteriorates, and the structural integrity of the cartilage progressively erodes. This compromised matrix not only reduces the cartilage's ability to support joint movement but also diminishes its shock-absorbing properties, contributing to the overall pathology of OA [15].

5.5 Inflammatory processes

In addition to mechanical stress and structural changes, inflammation emerges as a crucial driver of articular cartilage degeneration in the context of osteoarthritis (OA). The presence of pro-inflammatory cytokines, such as interleukin-1 (IL-1) and tumor necrosis factor-alpha (TNF-α), serves as a key trigger for the initiation and propagation of inflammatory processes within the joint microenvironment. When pro-inflammatory cytokines are present, chondrocytes undergo a shift in their gene expression profiles, resulting in an elevated production of catabolic enzymes, particularly matrix metalloproteinases (MMPs) and aggrecanases. These enzymes wield the ability to cleave vital components of the extracellular matrix. The upregulation of catabolic enzymes contributes to the accelerated breakdown of matrix components, while the synthesis of matrix molecules, including collagen and proteoglycans, is compromised. This disequilibrium skews the cartilage microenvironment toward degradation, intensifying the degenerative processes at play in OA. Furthermore, the inflammatory milieu can trigger the release of pro-inflammatory mediators, amplifying the inflammatory cascade and leading to a self-perpetuating loop of cartilage breakdown [46].

5.6 Fibrillation and erosion

Notably, one of the hallmark changes observed in the course of OA is the progression of cartilage surface fibrillation [47–50]. Fibrillation is characterized by the emergence of fissures and cracks on the cartilage surface, imparting a rough and irregular appearance [47]. This process of fibrillation initiates as a result of the cumulative wear and tear experienced by the cartilage during joint movement and loading [47, 51]. The constant exposure to mechanical stresses, particularly under the conditions of OA, gradually weakens the cartilage matrix and disrupts its integrity [51]. The weakened areas become prone to the formation of fissures, which gradually extend and deepen, ultimately resulting in more significant erosions [47, 51]. Over time, these erosions can progress through the layers of cartilage, eventually reaching the underlying subchondral bone [48, 52].

The consequences of cartilage fibrillation are multifaceted and impactful. The erosion of the once-smooth cartilage surface disturbs the normal mechanics of joint movement [52]. The smooth articulation between bones is compromised, leading to increased friction, uneven loading, and altered joint biomechanics [50]. This disruption in joint mechanics not only triggers pain but also exacerbates inflammation within the joint [52]. The exposed subchondral bone, lacking the protective cushioning of cartilage, becomes vulnerable to microtrauma and contributes to the inflammatory response [47, 52].

Moreover, the loss of the smooth cartilage surface compounds the mechanical stress placed on the remaining healthy cartilage [53]. The altered joint mechanics and the absence of the protective cartilage buffer force the remaining cartilage to bear an increased load, further accelerating its degeneration [47, 53]. As a result, a detrimental feedback loop ensues, where cartilage fibrillation and degradation drive joint pain, inflammation, and greater mechanical stress on the already compromised joint structures [47, 54].

5.7 Impact on synovium and joint capsule

In osteoarthritis (OA), the impact extends beyond articular cartilage degeneration to involve other joint structures, including the synovium—the lining of the joint capsule [46, 48]. The synovium plays a crucial role in joint lubrication, nutrient supply, and the regulation of inflammatory processes [46]. However, in OA, the synovium undergoes pathological changes that contribute to the disease progression [48]. One significant alteration observed in OA is synovial inflammation and thickening [55]. This inflammatory response in the synovium results in the release of pro-inflammatory cytokines, such as interleukin-1 (IL-1) and tumor necrosis factor-alpha (TNF-α), as well as various matrix-degrading enzymes. These cytokines and enzymes have detrimental effects on the joint environment fueling overall joint inflammation. The increased levels of inflammatory cytokines amplify the activity of chondrocytes that produce matrix-degrading enzymes, such as matrix metalloproteinases (MMPs), which contribute to cartilage degradation [46, 55, 56].

The interaction between synovial inflammation and articular cartilage degeneration emphasizes the systemic nature of OA and highlights the importance of considering the entire joint environment in disease management [55, 57]. Strategies aimed at mitigating OA progression need to address not only cartilage health but also the underlying inflammatory processes occurring within the synovium [46, 48, 55, 57]. By targeting synovial inflammation and its associated cytokines and enzymes, it may be possible to modulate the degenerative cascade and alleviate the overall burden of OA on joint health.

In conclusion, articular cartilage degeneration is a central feature of osteoarthritis. The complex interplay of mechanical stress, proteoglycan loss, collagen disorganization, chondrocyte dysfunction, and inflammatory processes contributes to the progressive breakdown of cartilage. Understanding these complex mechanisms is essential for developing targeted interventions to slow or halt the degenerative process and improve joint health in individuals affected by OA.

6. Subchondral bone changes and osteophyte formation

Subchondral bone plays a crucial role in supporting the articular cartilage and maintaining joint integrity. In osteoarthritis (OA), subchondral bone undergoes significant changes that contribute to the disease's progression and symptoms. This section explores the alterations in subchondral bone and the formation of osteophytes—bone spurs—associated with OA.

6.1 Subchondral bone changes in OA

In the context of osteoarthritis (OA), subchondral bone undergoes a series of complex structural and metabolic changes that significantly impact joint health and function. These changes are intimately linked to the altered mechanical environment resulting from the degeneration of articular cartilage, and they play a crucial role in the progression of the disease. These subchondral bone changes are not only a consequence of OA but also contribute to the overall pathophysiology of the condition. Some key subchondral bone changes include:

Sclerosis: One of the primary subchondral bone changes observed in OA is sclerosis. Subchondral bone sclerosis entails an increase in bone density, resulting in the thickening and hardening of the bone beneath the cartilage layer. This heightened density is believed to be a response to the heightened mechanical stress placed on the bone due to the loss of the protective cartilage cushioning [52, 58]. As the cartilage deteriorates, the subchondral bone attempts to compensate the diminished shock-absorbing capacity, leading to increased bone density. However, while this may initially serve as a protective mechanism, the excessive density can lead to altered joint mechanics and exacerbate the biomechanical imbalance that characterizes OA.

Cyst formation: Cyst formation is another notable alteration within the subchondral bone in OA. These cyst-like spaces develop as a result of bone resorption and subsequent remodeling processes. The presence of these cysts can weaken the structural integrity of the subchondral bone and contribute to joint pain and instability [58, 59]. The compromised bone structure further exacerbates the mechanical stresses experienced by the joint, creating a feedback loop that perpetuates the OA process.

Subchondral bone remodeling: Furthermore, OA leads to substantial changes in the process of subchondral bone remodeling. An imbalance emerges between the activities of osteoclasts, which are responsible for bone resorption, and osteoblasts, responsible for bone formation. This imbalance results in altered bone turnover, affecting the quality and structure of the bone tissue [58, 59]. The disrupted bone remodeling process contributes to the overall weakening of the subchondral bone, rendering it less capable of providing the necessary support for the articular cartilage above.

Collectively, these subchondral bone changes in OA not only reflect the adaptation of the bone to the altered joint mechanics but also play a significant role in driving the progression of the disease. The interconnected relationship between subchondral bone and articular cartilage highlights the complexity of OA's pathogenesis and underscores the importance of considering both tissue types when developing interventions to address this debilitating condition. The structural and metabolic changes within the subchondral bone add to the intricate web of factors that contribute to the overall deterioration of the joint in OA.

6.2 Osteophyte formation

Osteophytes, commonly referred to as bone spurs, represent a significant phenomenon within the context of osteoarthritis (OA), contributing to the complex interplay between joint degeneration and adaptive responses. These bony outgrowths develop at the margins of the affected joint and serve as a unique response of the body to the underlying pathological changes. Osteophytes play a dual role, attempting to stabilize the compromised joint while also potentially exacerbating joint symptoms and limitations [60, 61].

Etiology: The etiology of osteophyte formation involves a multifaceted interplay of factors. Mechanical stress on the subchondral bone, a direct consequence of cartilage breakdown, plays a pivotal role. Additionally, the inflammatory environment characteristic of OA contributes to osteophyte development. Inflammation triggers the release of growth factors that stimulate local bone-forming cells, or osteoblasts, to initiate the process of new bone formation [60]. This collaboration between mechanical stress, inflammation, and growth factors culminates in the gradual generation of osteophytes at the joint margins.

Radiographic appearance: Osteophytes are visible on X-rays and are a characteristic finding in OA. They appear as bony projections at the joint margins, often described as a "lipping" or "spurring" appearance [60].

Impact on joint function: While osteophytes serve a structural purpose by attempting to stabilize the joint, they can also introduce complications for the affected individual. The presence of osteophytes within the joint space can contribute to joint pain, stiffness, and a limited range of motion. In particular, larger osteophytes have the potential to impinge on adjacent soft tissues or compress nerves, further exacerbating pain and discomfort [61]. This impact on joint function can hinder daily activities and contribute to the overall burden of OA on an individual's quality of life.

In summary, osteophytes are a remarkable manifestation of the body's response to the challenges posed by OA. While they serve as an adaptive mechanism to stabilize the joint, their presence can also give rise to additional symptoms and functional limitations. The formation of osteophytes underlines the intricate relationship between joint degeneration and the body's attempts at compensation, further emphasizing the complexity of OA's pathophysiology.

6.3 Relationship with articular cartilage degeneration

Subchondral bone changes and osteophyte formation are closely linked with the degeneration of articular cartilage [58, 59]. The loss of cartilage results in increased mechanical stress on the subchondral bone, leading to sclerosis and cyst formation [59]. Osteophytes develop as a compensatory mechanism to stabilize the joint and redistribute forces [60, 61].

6.4 Clinical implications

The clinical implications of the interrelationship between subchondral bone changes, osteophyte formation, and articular cartilage degeneration are profound in the management of osteoarthritis (OA). These interconnected processes hold diagnostic and monitoring value, aiding healthcare professionals in accurately assessing disease progression through imaging techniques. Subchondral bone alterations and osteophyte presence can serve as indicators of disease severity, guiding treatment decisions and enabling predictions about the trajectory of OA. Treatment strategies can be refined to address both cartilage degeneration and subchondral bone abnormalities, leading to more effective interventions.

7. Emerging therapies and regenerative medicine for cartilage repair

Cartilage repair remains a significant challenge in the management of osteoarthritis (OA) and other cartilage-related conditions. Traditional treatments focus on symptom management and delaying disease progression, but they often fall short in promoting true cartilage regeneration. However, advancements in regenerative medicine offer promising approaches to stimulate cartilage repair and restoration. This section explores some of the emerging therapies and regenerative medicine techniques for cartilage repair.

7.1 Cell-based therapies

In recent years, cell-based therapies have gained prominence as potential solutions for cartilage repair. Progress in cell sourcing, scaffold design, and clinical implementation highlights the dynamic nature of this field. Despite challenges, including long-term efficacy and standardization, cell-based therapies remain at the forefront of cartilage repair strategies. Their transformative potential might hold the key to redefining the treatment landscape for cartilage-related conditions [62].

- *Autologous chondrocyte implantation (ACI)*: ACI involves the harvest and culture of a patient's own healthy chondrocytes. These cultured chondrocytes are then implanted into the damaged area of the cartilage, promoting tissue repair. ACI is suitable for larger cartilage defects and has shown encouraging outcomes in clinical trials [62, 63].

- *Matrix-induced autologous chondrocyte implantation (MACI)*: MACI is a variation of ACI that involves the use of a biodegradable scaffold to support the implanted chondrocytes. The scaffold enhances cell adhesion and proliferation, aiding in cartilage repair [64].

- Mesenchymal stem cell (MSC) therapy: MSCs are multipotent cells with the potential to differentiate into various cell types, including chondrocytes [65]. MSC therapy involves the injection of MSCs into the damaged cartilage, promoting its regeneration and repair. MSCs can be derived from various sources, such as bone marrow, adipose tissue, and umbilical cord tissue [65, 66].

7.2 Tissue engineering

Tissue engineering, a multidisciplinary field at the intersection of biology, engineering, and medicine, has emerged as a groundbreaking approach to address the limitations of conventional medical treatments and organ transplantation. In orthopedics, it is revolutionizing joint and bone repair; in cardiology, it is advancing heart tissue regeneration. Skin substitutes aid wound healing, while engineered organs strive to overcome the organ shortage crisis. Tissue engineering has emerged as a transformative solution for cartilage repair, aiming to overcome the challenges posed by the limited self-renewal capacity of articular cartilage [67].

- *3D bioprinting*: 3D bioprinting allows the creation of custom-designed scaffolds that mimic the complex architecture of native cartilage. Chondrocytes or MSCs are seeded onto these scaffolds, which are then implanted into the defect site to promote cartilage regeneration [68, 69].

- *Decellularized matrix-based approaches*: Decellularized cartilage matrix provides an excellent biological scaffold for cartilage repair. The decellularized matrix retains the natural tissue architecture and biochemical cues necessary for cartilage regeneration. Cells, such as chondrocytes or MSCs, are then seeded onto the scaffold for implantation [69, 70].

7.3 Growth factor and cytokine therapy

Growth factors and cytokines play pivotal roles in orchestrating cellular responses, driving tissue repair, and modulating inflammatory processes. In the realm of cartilage repair, harnessing the potential of these signaling molecules has emerged as a promising strategy. Growth factors such as transforming growth factor-beta (TGF-β), fibroblast growth factors (FGFs), and insulin-like growth factors (IGFs) exert profound effects on chondrocytes and stem cells, regulating proliferation, differentiation, and extracellular matrix synthesis [71]. Direct delivery of growth factors or cytokines to the injury site aims to enhance the local environment for endogenous repair [71, 72]. Despite the promise, growth factor and cytokine therapies face challenges. Achieving optimal dosing, ensuring sustained release, and avoiding unwanted effects are hurdles. Precise spatial and temporal control of signaling activation is necessary to avoid undesirable outcomes, such as hypertrophy or fibrosis [71, 73].

- *Platelet-rich plasma (PRP)*: PRP is extracted from the patient's own blood and contains a high concentration of growth factors. When injected into the damaged cartilage, PRP can stimulate chondrocyte proliferation and ECM synthesis, promoting cartilage repair. Scientific studies have shown that PRP's application can lead to improvements in cartilage repair and pain reduction. While its efficacy can vary based on factors such as the severity of the cartilage damage and the patient's overall health, PRP offers a minimally invasive approach to stimulate the body's natural healing mechanisms and support the restoration of cartilage tissue [74–76].

- Cytokine therapy: Specific cytokines, such as transforming growth factor-beta (TGF-β) and insulin-like growth factor (IGF), have shown potential in promoting chondrogenesis and cartilage regeneration. They can be delivered directly to the damaged cartilage to enhance repair processes [33].

7.4 Gene therapy

Gene therapy, a revolutionary approach to manipulate genetic information, has appeared as a promising frontier in the realm of cartilage repair [77]. Gene therapy encompasses the delivery of therapeutic genes to target cells, either by viral vectors or by nonviral methods [78]. The introduced genes can encode growth factors, cytokines, or molecular regulators that modulate cellular responses, foster chondrogenesis, and promote tissue regeneration. In osteoarthritis, genes encoding anti-inflammatory factors can mitigate inflammation-driven cartilage degradation [77, 78]. Gene therapy faces challenges, including the achievement of sustainable and localized gene expression. Off-target effects and immune responses are critical safety concerns. Balancing therapeutic efficacy with safety necessitates meticulous design and testing [79]. As research advances, gene therapy's potential to revolutionize cartilage repair is increasingly evident, paving the way for a new era of targeted, genetically tailored regenerative treatments.

7.5 Nanotechnology

Nanotechnology has revolutionized biomaterial design. Nanostructured scaffolds, inspired by cartilage's extracellular matrix, offer precise control over mechanical properties and cell-material interactions. These scaffolds mimic native tissue architecture, enhancing cellular adhesion, proliferation, and chondrogenic differentiation [80]. Nanoparticles serve as carriers for precise and targeted delivery of growth factors and therapeutic agents. By encapsulating these molecules, nanotechnology enables sustained release, thus enhancing their bioavailability and therapeutic efficacy. This approach minimizes off-target effects and optimizes tissue healing. Nanotechnology extends beyond repair to diagnostics and monitoring. Nanosensors and imaging agents offer real-time insights into tissue health, enabling personalized treatment strategies and timely interventions [80, 81]. Nanotechnology's immense potential is accompanied with challenges. Ensuring biocompatibility, long-term safety, and scalable manufacturing is a critical concern. Additionally, ethical and regulatory considerations arise as nanotechnology bridges the divide between medicine and material science [81].

7.6 Exosome therapy

Exosomes are carriers of bioactive molecules, including proteins, nucleic acids, and lipids, functioning as messengers between cells. Their unique ability to modulate recipient cell behavior makes them intriguing candidates for therapeutic interventions [82]. In cartilage repair, exosomes play roles in anti-inflammatory responses, extracellular matrix remodeling, and promotion of chondrogenic differentiation. Exosome therapy involves isolating and purifying exosomes from various sources, including mesenchymal stem cells (MSCs) and chondrocytes. Techniques like ultracentrifugation and precipitation are commonly employed, ensuring the enrichment of exosomal cargo. Direct exosome administration promotes chondroprotection, reducing inflammation and inhibiting cartilage degradation [83]. Encapsulation of exosomes in biomaterial scaffolds augments their release profile, enhancing their therapeutic effects [82, 83]. Exosome therapy circumvents challenges associated with cell-based therapies, such as immune responses and potential tumorigenicity. However, standardized isolation methods, cargo loading, and dosage optimization pose challenges [82].

While these emerging therapies show great promise for cartilage repair, further research is needed to optimize their effectiveness and safety. Additionally, research trials are being conducted to evaluate the long-term outcomes and potential side effects of these novel approaches. As research continues, regenerative medicine holds incredible potential in revolutionizing the treatment landscape for cartilage repair and improving the quality of life for persons with cartilage-related conditions.

8. Challenges and potential solutions in developing cartilage-targeted treatments

Developing effective and targeted treatments for cartilage-related conditions, such as osteoarthritis (OA), presents various challenges due to the intricate nature of cartilage tissue and its unique features. Overcoming these challenges is crucial to improving patient outcomes and achieving successful cartilage repair and regeneration. Here are various fundamental challenges and potential solutions in formulating cartilage-targeted treatments:

8.1 Limited blood supply and innervation

- *Challenge*: Cartilage is avascular and lacks nerve fibers, making it difficult to deliver drugs and therapeutic agents directly to the affected area.
- *Solution*: Developing targeted drug delivery systems, such as nanoparticles or hydrogels, can help deliver medications precisely to the cartilage site. Intra-articular injections and tissue engineering approaches can also be used for localized delivery [84–86].

8.2 Low cellular activity

- *Challenge*: Chondrocytes, the cells within cartilage, have low metabolic activity, limiting their capacity of self-repair and regeneration.
- *Solution*: Stimulating chondrocyte activity through growth factors, gene therapy, or cell-based therapies can enhance cartilage repair and regeneration [85].

8.3 Complex cartilage structure

- *Challenge*: Cartilage has a unique extracellular matrix (ECM) composition, making it challenging to replicate *in vitro* for tissue engineering approaches.
- *Solution*: Decellularized matrix-based approaches that retain the natural ECM structure, combined with cell-based therapies, can enhance cartilage repair [85].

8.4 Inflammatory microenvironment

- *Challenge:* Inflammatory processes in OA can inhibit cartilage repair and regeneration.
- *Solution:* Targeting inflammation through the use of anti-inflammatory drugs, cytokine therapy, or regenerative agents can promote a more favorable microenvironment for cartilage repair [16, 19].

8.5 Disease heterogeneity

- *Challenge:* OA is a heterogeneous condition with varying degrees of cartilage damage and different underlying mechanisms.

- *Solution:* Personalized medicine approaches, such as identifying specific biomarkers and disease phenotypes, can help tailor treatments to individual patients [87].

8.6 Integration with native cartilage

- *Challenge:* Ensuring that the repaired or regenerated cartilage integrates seamlessly with the native cartilage is crucial for long-term success.

- *Solution:* Optimization of tissue engineering approaches, such as scaffold design and cell source selection, can enhance integration and biomechanical properties of the repaired tissue [88].

8.7 Long-term efficacy and safety

- *Challenge:* Ensuring that cartilage-targeted treatments have durable efficacy and safety profiles is critical for their clinical use.

- *Solution:* Conducting robust preclinical and clinical research trials to estimate the long-term outcomes and safety of these treatments is essential [89].

8.8 High costs and accessibility

- *Challenge:* Some advanced cartilage-targeted treatments may be costly and not readily accessible to all patients.

- *Solution:* Continued research and development may lead to more cost-effective and scalable treatments, making them more widely available [90].

In conclusion, developing effective cartilage-targeted treatments requires addressing the unique challenges presented by cartilage tissue. By leveraging advancements in drug delivery, tissue engineering, regenerative medicine, and personalized medicine, potential solutions can be found to improve cartilage repair, regeneration, and overall patient outcomes in cartilage-related conditions like osteoarthritis.

9. Call for continued research and collaboration in the field of cartilage and osteoarthritis

The study of cartilage and its related conditions, particularly osteoarthritis (OA), is of utmost importance in the field of orthopedics and musculoskeletal research. As we strive to improve patient outcomes and find effective treatments for cartilage-related disorders, a call for continued research and collaboration becomes imperative. By fostering a collaborative and multidisciplinary approach, we can unlock new insights, address current challenges, and develop innovative therapies for cartilage repair, regeneration, and osteoarthritis management.

- *Advancing basic science knowledge*: A deeper understanding of cartilage biology, molecular signaling pathways, and the intricate interactions between inflammatory mediators and cartilage cells is essential. Basic science research will pave the way for the advancement of targeted therapies and interventions.

- *Exploring regenerative medicine*: Regenerative medicine offers great promise in the realm of cartilage repair and regeneration. Continued research in stem cell biology, tissue engineering, and gene therapy will enable us to harness the regenerative potential of the body for cartilage healing.

- Investigating biomarkers and disease phenotypes: Identifying reliable biomarkers for cartilage degeneration and osteoarthritis progression can aid in early diagnosis, prognostication, and monitoring treatment response. Additionally, studying disease phenotypes will help tailor personalized medicine approaches for improved patient care.

- *Enhancing drug delivery systems:* Developing effective drug delivery systems for targeted cartilage treatment remains a challenge. Collaborative efforts between pharmacologists, engineers, and clinicians can lead to innovative solutions that deliver therapeutic agents precisely to the affected cartilage site.

- *Conducting longitudinal studies:* Longitudinal studies tracking patients over extended periods are crucial to understanding the natural progression of osteoarthritis and evaluating the long-term efficacy and safety of different treatments.

10. Conclusion

The journey through the intricate web of osteoarthritis (OA) pathogenesis has illuminated the multifaceted interplay of cellular, molecular, and biomechanical factors that orchestrate the degeneration of articular cartilage. OA, once considered a wear-and-tear phenomenon, has revealed itself as a dynamic disease driven by a plethora of intricate mechanisms.

The avascular and aneural nature of cartilage poses a challenge in delivering targeted therapies directly to the affected area. However, as highlighted in this chapter, innovative solutions are emerging. Nanoparticles, hydrogels, and tissue engineering approaches offer avenues for precise drug delivery, offering hope for localized treatment strategies that could potentially alter the trajectory of OA progression. Moreover, insights into the inflammatory microenvironment and cellular signaling have paved the way for novel interventions that target inflammation, reshaping the local milieu to support repair rather than degradation.

As our understanding of OA deepens, the significance of individualized approaches becomes paramount. Disease heterogeneity stresses the need for personalized medicine, harnessing the power of biomarkers and phenotypic characterizations to tailor treatments to patients' specific needs. Furthermore, the imperative to seamlessly integrate repaired or regenerated cartilage with native tissue highlights the importance of scaffold design and cell selection in tissue engineering strategies, ensuring long-term success.

Yet challenges remain from ensuring long-term efficacy and safety to addressing the accessibility of advanced treatments. With each challenge, however, comes opportunity. Continued research and clinical trials hold the promise of refining and validating interventions, making them more accessible, affordable, and effective for a broader spectrum of patients.

In the grand narrative of cartilage degeneration and osteoarthritis pathogenesis, this chapter is but a stepping stone. As the field continues to evolve, forging connections between cellular mechanisms, biomaterials, and clinical practice, the journey to mitigate the impact of osteoarthritis takes on new dimensions. By understanding the intricate web of factors that contribute to OA, we will be empowered to envision a future where its progression is not inevitable, but rather a challenge that can be met with innovation, compassion, and scientific rigor.

Acknowledgements

I would like to express my deepest gratitude to my beloved wife "Samia Khurshid," whose unwavering support, encouragement, and understanding have been the cornerstone of my journey in bringing this chapter to fruition. Her patience and constant belief in my abilities have been a wellspring of inspiration. Her presence has provided the calm amidst the storms of research and writing, and her love has been the guiding light that illuminated my path. This work would not have been possible without her love, sacrifices, and unwavering belief in my aspirations.

Dr. Saif ur Rehman

Conflict of interest

The authors declare no conflict of interest.

Author details

Saif ur Rehman[1*], Safdar Iqbal[2], Muhammad Umair Shahid[1],
Muhammad Soman Jahangir[3] and Adnan Latif Malik[1]

1 Sir Ganga Ram Hospital, Lahore, Pakistan

2 Jubail Medicare Complex, Jubail, KSA

3 Bahawal Victoria Hospital, Bahawalpur, Pakistan

*Address all correspondence to: drsrehman@outlook.com

References

[1] Pirozzi KM. Histophysiology of fibrocartilage. Clinics in Podiatric Medicine and Surgery. 2022;**39**(3):363-370. DOI: 10.1016/j.cpm.2022.02.002

[2] Gahunia HK, Pritzker KPH. Structure and function of articular cartilage. In: Gahunia HK, Gross AE, Pritzker KPH, Babyn PS, Murnaghan L, editors. Articular Cartilage of the Knee Health, Disease and Therapy. Springer Nature; 2020. pp. 36-104

[3] Sophia Fox AJ, Bedi A, Rodeo SA. The basic science of articular cartilage: Structure, composition, and function. Sports Health. 2009;**1**(6):461-468. DOI: 10.1177/1941738109350438

[4] Żylińska B, Sobczyńska-Rak A, Lisiecka U, Stodolak-Zych E, Jarosz Ł, Szponder T. Structure and pathologies of articular cartilage. In Vivo. 2021;**35**(3):1355-1363. DOI: 10.21873/invivo.12388

[5] Vincent TL, McClurg O, Troeberg L. The extracellular matrix of articular cartilage controls the bioavailability of pericellular matrix-bound growth factors to drive tissue homeostasis and repair. International Journal of Molecular Sciences. 2022;**23**(11):6003. DOI: 10.3390/ijms23116003

[6] Verzijl N, DeGroot J, Thorpe SR, Bank RA, Shaw JN, Lyons TJ, et al. Effect of collagen turnover on the accumulation of advanced glycation end products. The Journal of Biological Chemistry. 2000;**275**:39027-39031

[7] Mostakhdemin M, Nand A, Ramezani M. Articular and artificial cartilage, characteristics, properties and testing approaches—A review. Polymers (Basel). 2021;**13**(12):2000. DOI: 10.3390/polym13122000

[8] Iozzo RV, Schaefer L. Proteoglycan form and function: A comprehensive nomenclature of proteoglycans. Matrix Biology. 2015;**42**:11-55. DOI: 10.1016/j.matbio.2015.02.003

[9] Bhosale AM, Richardson JB. Articular cartilage: Structure, injuries and review of management. British Medical Bulletin. 2008;**87**:77-95. DOI: 10.1093/bmb/ldn025

[10] Marczyński W. Pathology of articular cartilage – Dynamics of changes, prevention. Wiadomości Lekarskie. 2007;**60**(1):53-59

[11] Youn I, Choi JB, Cao L, Setton LA, Guilak F. Zonal variations in the three-dimensional morphology of the chondron measured in situ using confocal microscopy. Osteoarthritis and Cartilage. 2006;**14**(9):889-897. DOI: 10.1016/j.joca.2006.02.017

[12] Pearle AD, Warren RF, Rodeo SA. Basic science of articular cartilage and osteoarthritis. Clinics in Sports Medicine. 2005;**24**(1):1-12. DOI: 10.1016/j.csm.2004.08.007

[13] Decker RS, Koyama E, Pacifici M. Articular cartilage: Structural and developmental intricacies and questions. Current Osteoporosis Reports. 2015;**13**(6):407-414. DOI: 10.1007/s11914-015-0290-z

[14] Eschweiler J, Horn N, Rath B, Betsch M, Baroncini A, Tingart M, et al. The biomechanics of cartilage—An overview. Life (Basel). 2021;**11**(4):302. DOI: 10.3390/life11040302

[15] Goldring MB, Goldring SR. Articular cartilage and subchondral bone in

the pathogenesis of osteoarthritis. Annals of the New York Academy of Sciences. 2010;**1192**:230-237. DOI: 10.1111/j.1749-6632.2009.05240.x

[16] Lotz M, Loeser RF. Effects of aging on articular cartilage homeostasis. Bone. 2012;**51**(2):241-248. DOI: 10.1016/j.bone.2012.03.023

[17] Brittberg M, Peterson L, Sjögren-Jansson E, Tallheden T, Lindahl A. Articular cartilage engineering with autologous chondrocyte transplantation. A review of recent developments. The Journal of Bone and Joint Surgery. American Volume. 2003;**85-A**(Suppl-3):109-115. DOI: 10.2106/00004623-200300003-00017

[18] Armiento AR, Alini M, Stoddart MJ. Articular fibrocartilage—Why does hyaline cartilage fail to repair? Advanced Drug Delivery Reviews. 2019;**146**:289-305. DOI: 10.1016/j.addr.2018.12.015

[19] Loeser RF, Goldring SR, Scanzello CR, Goldring MB. Osteoarthritis: A disease of the joint as an organ. Arthritis and Rheumatism. 2012;**64**(6):1697-1707. DOI: 10.1002/art.34453

[20] Masson AO, Krawetz RJ. Understanding cartilage protection in OA and injury: A spectrum of possibilities. BMC Musculoskeletal Disorders. 2020;**21**(1):432. DOI: 10.1186/s12891-020-03363-6

[21] Gross AE, Shasha N, Aubin P. Long-term followup of the use of fresh osteochondral allografts for posttraumatic knee defects. Clinical Orthopaedics and Related Research. 2005;**435**:79-87

[22] Madry H, Orth P, Cucchiarini M. Role of the subchondral bone in articular cartilage degeneration and repair. The Journal of the American Academy of Orthopaedic Surgeons. 2016;**24**(4):e45-e46. DOI: 10.5435/JAAOS-D-16-00096

[23] Salter RB, Simmonds DF, Malcolm BW, Rumble EJ, MacMichael D, Clements ND. The biological effect of continuous passive motion on the healing of full-thickness defects in articular cartilage. An experimental investigation in the rabbit. The Journal of Bone and Joint Surgery. American Volume. 1980;**62**(8):1232-1251

[24] Fahy N, Alini M, Stoddart MJ. Mechanical stimulation of mesenchymal stem cells: Implications for cartilage tissue engineering. Journal of Orthopaedic Research. 2018;**36**(1):52-63. DOI: 10.1002/jor.23670

[25] Iseki T, Rothrauff BB, Kihara S, Sasaki H, Yoshiya S, Fu FH, et al. Dynamic compressive loading improves cartilage repair in an in vitro model of microfracture: Comparison of 2 mechanical loading regimens on simulated microfracture based on fibrin gel scaffolds encapsulating connective tissue progenitor cells. The American Journal of Sports Medicine. 2019;**47**(9):2188-2199. DOI: 10.1177/0363546519855645

[26] Sanchez-Adams J, Leddy HA, McNulty AL, O'Conor CJ, Guilak F. The mechanobiology of articular cartilage: Bearing the burden of osteoarthritis. Current Rheumatology Reports. 2014;**16**(10):451. DOI: 10.1007/s11926-014-0451-6

[27] Lohmander LS, Englund PM, Dahl LL, Roos EM. The long-term consequence of anterior cruciate ligament and meniscus injuries: Osteoarthritis. The American Journal of Sports Medicine. 2007;**35**(10):1756-1769. DOI: 10.1177/0363546507307396

[28] Guo Q, Wang Y, Xu D, Nossent J, Pavlos NJ, Xu J. Rheumatoid arthritis: Pathological mechanisms and modern pharmacologic therapies. Bone Research. 2018;**6**:15. DOI: 10.1038/s41413-018-0016-9

[29] Coutant F, Miossec P. Evolving concepts of the pathogenesis of rheumatoid arthritis with focus on the early and late stages. Current Opinion in Rheumatology. 2020;**32**(1):57-63. DOI: 10.1097/BOR.0000000000000664

[30] Chawla S, Mainardi A, Majumder N, Dönges L, Kumar B, Occhetta P, et al. Chondrocyte hypertrophy in osteoarthritis: Mechanistic studies and models for the identification of new therapeutic strategies. Cell. 2022;**11**(24):4034. DOI: 10.3390/cells11244034

[31] Goldring MB. The role of the chondrocyte in osteoarthritis. Arthritis and Rheumatism. 2000;**43**(9):1916-1926. DOI: 10.1002/1529-0131(200009)43:9<1916::AID-ANR2>3.0.CO;2-I

[32] Molnar V, Matišić V, Kodvanj I, Bjelica R, Jeleč Ž, Hudetz D, et al. Cytokines and chemokines involved in osteoarthritis pathogenesis. International Journal of Molecular Sciences. 2021;**22**(17):9208. DOI: 10.3390/ijms22179208

[33] Wen C, Xu L, Xu X, Wang D, Liang Y, Duan L. Insulin-like growth factor-1 in articular cartilage repair for osteoarthritis treatment. Arthritis Research & Therapy. 2021;**23**(1):277. DOI: 10.1186/s13075-021-02662-0

[34] Migliorini F, Eschweiler J, Schenker H, Baroncini A, Tingart M, Maffulli N. Surgical management of focal chondral defects of the knee: A Bayesian network meta-analysis. Journal of Orthopaedic Surgery and Research. 2021;**16**(1):543. DOI: 10.1186/s13018-021-02684-z

[35] Rehman SU, Afzal T, Iqbal S, Zain-Ul-Abidin M, Aslam F. Comparison of disease progression in patients with osteoarthritis of knee treated with versus without intraarticular injection of corticosteroids. Pakistan Journal of Medical & Health Sciences. 2022;**16**(2):255-257. DOI: 10.53350/pjmhs22162255

[36] Zeng C, Lane NE, Hunter DJ, Wei J, Choi HK, McAlindon TE, et al. Intra-articular corticosteroids and the risk of knee osteoarthritis progression: Results from the osteoarthritis initiative. Osteoarthritis and Cartilage. 2019;**27**(6):855-862. DOI: 10.1016/j.joca.2019.01.007

[37] Arner JW, Irvine JN, Zheng L, Gale T, Thorhauer E, Hankins M, et al. The effects of anterior cruciate ligament deficiency on the meniscus and articular cartilage: A novel dynamic in vitro pilot study. Orthopaedic Journal of Sports Medicine. 2016;**4**(4):2325967116639895. DOI: 10.1177/2325967116639895

[38] Sutter EG, Liu B, Utturkar GM, Widmyer MR, Spritzer CE, Cutcliffe HC, et al. Effects of anterior cruciate ligament deficiency on tibiofemoral cartilage thickness and strains in response to hopping. The American Journal of Sports Medicine. 2019;**47**(1):96-103. DOI: 10.1177/0363546518802225

[39] Deckers C, Stephan P, Wever KE, Hooijmans CR, Hannink G. The protective effect of anterior cruciate ligament reconstruction on articular cartilage: A systematic review of animal studies. Osteoarthritis and Cartilage. 2019;**27**(2):219-229. DOI: 10.1016/j.joca.2018.10.001

[40] Anderson DD, Chubinskaya S, Guilak F, Martin JA, Oegema TR,

Olson SA, et al. Post-traumatic osteoarthritis: Improved understanding and opportunities for early intervention. Journal of Orthopaedic Research. 2011;**29**(6):802-809. DOI: 10.1002/jor.21359

[41] Peffers M, Liu X, Clegg P. Transcriptomic signatures in cartilage ageing. Arthritis Research & Therapy. 2013;**15**(4):R98. DOI: 10.1186/ar4278

[42] Guilak F, Fermor B, Keefe FJ, Kraus VB, Olson SA, Pisetsky DS, et al. The role of biomechanics and inflammation in cartilage injury and repair. Clinical Orthopaedics and Related Research. 2004;**423**:17-26. DOI: 10.1097/01.blo.0000131233.83640.91

[43] Roughley PJ, Mort JS. The role of aggrecan in normal and osteoarthritic cartilage. Journal of Experimental Orthopaedics. 2014;**1**(1):8. DOI: 10.1186/s40634-014-0008-7

[44] Eyre D. Collagen of articular cartilage. Arthritis Research. 2002;**4**(1):30-35. DOI: 10.1186/ar380

[45] Chen H, Tan XN, Hu S, Liu RQ, Peng LH, Li YM, et al. Molecular mechanisms of chondrocyte proliferation and differentiation. Frontiers in Cell and Development Biology. 2021;**9**:664168. DOI: 10.3389/fcell.2021.664168

[46] Goldring MB, Otero M. Inflammation in osteoarthritis. Current Opinion in Rheumatology. 2011;**23**(5):471-478. DOI: 10.1097/BOR.0b013e328349c2b1

[47] Primorac D, Molnar V, Rod E, Jeleč Ž, Čukelj F, Matišić V, et al. Knee osteoarthritis: A review of pathogenesis and state-of-the-art non-operative therapeutic considerations. Genes (Basel). 2020;**11**(8):854. DOI: 10.3390/genes11080854

[48] Goldring MB, Goldring SR. Osteoarthritis. Journal of Cellular Physiology. 2007;**213**(3):626-634. DOI: 10.1002/jcp.21258

[49] Burr DB. Anatomy and physiology of the mineralized tissues: Role in the pathogenesis of osteoarthrosis. Osteoarthritis Cartilage. 2004;**12** (Suppl. A):S20-S30. DOI: 10.1016/j.joca.2003.09.016

[50] Buckwalter JA, Mankin HJ. Articular cartilage: Degeneration and osteoarthritis, repair, regeneration, and transplantation. Instructional Course Lectures. 1998;**47**:487-504

[51] Aigner T, Söder S, Gebhard PM, McAlinden A, Haag J. Mechanisms of disease: Role of chondrocytes in the pathogenesis of osteoarthritis--structure, chaos and senescence. Nature Clinical Practice. Rheumatology. 2007;**3**(7):391-399. DOI: 10.1038/ncprheum0534

[52] Lories RJ, Luyten FP. The bone-cartilage unit in osteoarthritis. Nature Reviews Rheumatology. 2011;7(1):43-49. DOI: 10.1038/nrrheum.2010.197

[53] Kuyinu EL, Narayanan G, Nair LS, Laurencin CT. Animal models of osteoarthritis: Classification, update, and measurement of outcomes. Journal of Orthopaedic Surgery and Research. 2016;**11**:19. DOI: 10.1186/s13018-016-0346-5

[54] Madry H, Luyten FP, Facchini A. Biological aspects of early osteoarthritis. Knee Surgery, Sports Traumatology, Arthroscopy. 2012;**20**(3):407-422. DOI: 10.1007/s00167-011-1705-8

[55] Scanzello CR, Goldring SR. The role of synovitis in osteoarthritis

pathogenesis. Bone. 2012;**51**(2):249-257. DOI: 10.1016/j.bone.2012.02.012

[56] Pearle AD, Scanzello CR, George S, Mandl LA, DiCarlo EF, Peterson M, et al. Elevated high-sensitivity C-reactive protein levels are associated with local inflammatory findings in patients with osteoarthritis. Osteoarthritis and Cartilage. 2007;**15**(5):516-523. DOI: 10.1016/j.joca.2006.10.010

[57] Roach HI, Aigner T, Kouri JB. Chondroptosis: A variant of apoptotic cell death in chondrocytes? Apoptosis. 2004;**9**(3):265-277. DOI: 10.1023/b:appt.0000025803.17498.26

[58] Goldring SR, Goldring MB. Bone and cartilage in osteoarthritis: Is what's best for one good or bad for the other? Arthritis Research & Therapy. 2010;**12**(5):143. DOI: 10.1186/ar3135

[59] Burr DB, Gallant MA. Bone remodelling in osteoarthritis. Nature Reviews Rheumatology. 2012;**8**(11):665-673. DOI: 10.1038/nrrheum.2012.130

[60] Martel-Pelletier J, Barr AJ, Cicuttini FM, Conaghan PG, Cooper C, Goldring MB, et al. Osteoarthritis. Nature Reviews. Disease Primers. 2016;**2**:16072. DOI: 10.1038/nrdp.2016.72

[61] Neogi T. The epidemiology and impact of pain in osteoarthritis. Osteoarthritis and Cartilage. 2013;**21**(9):1145-1153. DOI: 10.1016/j.joca.2013.03.018

[62] Jeuken RM, van Hugten PPW, Roth AK, Timur UT, Boymans TAEJ, van Rhijn LW, et al. A systematic review of focal cartilage defect treatments in middle-aged versus younger patients. Orthopaedic Journal of Sports Medicine. 2021;**9**(10):23259671211031244. DOI: 10.1177/23259671211031244

[63] Niemeyer P, Laute V, Zinser W, Becher C, Diehl P, Kolombe T, et al. Clinical outcome and success rates of ACI for cartilage defects of the patella: A subgroup analysis from a controlled randomized clinical phase II trial (CODIS study). Archives of Orthopaedic and Trauma Surgery. 2020;**140**(6):717-725. DOI: 10.1007/s00402-019-03264-x

[64] Brittberg M, Recker D, Ilgenfritz J, Saris DBF, SUMMIT Extension Study Group. Matrix-applied characterized autologous cultured chondrocytes versus microfracture: Five-year follow-up of a prospective randomized trial. The American Journal of Sports Medicine. 2018;**46**(6):1343-1351. DOI: 10.1177/0363546518756976

[65] Le H, Xu W, Zhuang X, Chang F, Wang Y, Ding J. Mesenchymal stem cells for cartilage regeneration. Journal of Tissue Engineering. 2020;**11**:2041731420943839. DOI: 10.1177/2041731420943839

[66] Arshi A, Petrigliano FA, Williams RJ, Jones KJ. Stem cell treatment for knee articular cartilage defects and osteoarthritis. Current Reviews in Musculoskeletal Medicine. 2020;**13**(1):20-27. DOI: 10.1007/s12178-020-09598-z

[67] Esdaille CJ, Washington KS, Laurencin CT. Regenerative engineering: A review of recent advances and future directions. Regenerative Medicine. 2021;**16**(5):495-512. DOI: 10.2217/rme-2021-0016

[68] Francis SL, Di Bella C, Wallace GG, Choong PFM. Cartilage tissue engineering using stem cells and bioprinting technology-barriers to clinical translation. Frontiers in Surgery. 2018;**5**:70. DOI: 10.3389/fsurg.2018.00070

[69] Sahranavard M, Sarkari S, Safavi S, Ghorbani F. Three-dimensional bio-printing of decellularized extracellular matrix-based bio-inks for cartilage regeneration: A systematic review. Biomaterials Translational. 2022;**3**(2):105-115. DOI: 10.12336/biomatertransl.2022.02.004

[70] Sun Y, Yan L, Chen S, Pei M. Functionality of decellularized matrix in cartilage regeneration: A comparison of tissue versus cell sources. Acta Biomaterialia. 2018;**74**:56-73. DOI: 10.1016/j.actbio.2018.04.048

[71] Goldring MB. Chondrogenesis, chondrocyte differentiation, and articular cartilage metabolism in health and osteoarthritis. Therapeutic Advances in Musculoskeletal Disease. 2012;**4**(4):269-285. DOI: 10.1177/1759720X12448454

[72] Kapoor M, Martel-Pelletier J, Lajeunesse D, Pelletier JP, Fahmi H. Role of proinflammatory cytokines in the pathophysiology of osteoarthritis. Nature Reviews Rheumatology. 2011;**7**(1):33-42. DOI: 10.1038/nrrheum.2010.196

[73] Bessa PC, Casal M, Reis RL. Bone morphogenetic proteins in tissue engineering: The road from laboratory to clinic, part II (BMP delivery). Journal of Tissue Engineering and Regenerative Medicine. 2008;**2**(2):81-96. DOI: 10.1002/term.74

[74] Filardo G, Previtali D, Napoli F, Candrian C, Zaffagnini S, Grassi A. PRP injections for the treatment of knee osteoarthritis: A meta-analysis of randomized controlled trials. Cartilage. 2021;**13**(Suppl. 1):364-375. DOI: 10.1177/1947603520931170

[75] Moretti L, Maccagnano G, Coviello M, Cassano GD, Franchini A, Laneve A, et al. Platelet rich plasma injections for knee osteoarthritis treatment: A prospective clinical study. Journal of Clinical Medicine. 2022;**11**(9):2640. DOI: 10.3390/jcm11092640

[76] Shahid A, Malik A, Bukhari A, Shaikh A, Rutherford J, Barkatali B. Do platelet-rich plasma injections for knee osteoarthritis work? Cureus. 2023;**15**(2):e34533. DOI: 10.7759/cureus.34533

[77] Li X, Shen L, Deng Z, Huang Z. New treatment for osteoarthritis: Gene therapy. Precision Clinical Medicine. 2023;**6**(2):pbad014. DOI: 10.1093/pcmedi/pbad014

[78] Uzieliene I, Kalvaityte U, Bernotiene E, Mobasheri A. Non-viral gene therapy for osteoarthritis. Frontiers in Bioengineering and Biotechnology. 2021;**8**:618399. DOI: 10.3389/fbioe.2020.618399

[79] Watson-Levings RS, Palmer GD, Levings PP, Dacanay EA, Evans CH, Ghivizzani SC. Gene therapy in orthopaedics: Progress and challenges in pre-clinical development and translation. Frontiers in Bioengineering and Biotechnology. 2022;**10**:901317. DOI: 10.3389/fbioe.2022.901317

[80] Qiao K, Xu L, Tang J, Wang Q, Lim KS, Hooper G, et al. The advances in nanomedicine for bone and cartilage repair. Journal of Nanobiotechnology. 2022;**20**(1):141. DOI: 10.1186/s12951-022-01342-8

[81] Eftekhari A, Maleki Dizaj S, Sharifi S, Salatin S, Rahbar Saadat Y, Zununi Vahed S, et al. The use of nanomaterials in tissue engineering for cartilage regeneration; current approaches and future perspectives. International Journal of Molecular

Sciences. 2020;**21**(2):536. DOI: 10.3390/ijms21020536

[82] Shao J, Zaro J, Shen Y. Advances in exosome-based drug delivery and tumor targeting: From tissue distribution to intracellular fate. International Journal of Nanomedicine. 2020;**15**:9355-9371. DOI: 10.2147/IJN.S281890

[83] Chen LY, Kao TW, Chen CC, Niaz N, Lee HL, Chen YH, et al. Frontier review of the molecular mechanisms and current approaches of stem cell-derived exosomes. Cell. 2023;**12**(7):1018. DOI: 10.3390/cells12071018

[84] Nguyen LH, Annabi N, Nikkhah M, Bae H, Binan L, Park S, et al. Vascularized bone tissue engineering: Approaches for potential improvement. Tissue Engineering. Part B, Reviews. 2012;**18**(5):363-382. DOI: 10.1089/ten.TEB.2012.0012

[85] Pati F, Song TH, Rijal G, Jang J, Kim SW, Cho DW. Ornamenting 3D printed scaffolds with cell-laid extracellular matrix for bone tissue regeneration. Biomaterials. 2015;**37**:230-241. DOI: 10.1016/j.biomaterials.2014.10.012

[86] Spiller KL, Maher SA, Lowman AM. Hydrogels for the repair of articular cartilage defects. Tissue Engineering. Part B, Reviews. 2011;**17**(4):281-299. DOI: 10.1089/ten.TEB.2011.0077

[87] Kraus VB, Blanco FJ, Englund M, Henrotin Y, Lohmander LS, Losina E, et al. OARSI clinical trials recommendations: Soluble biomarker assessments in clinical trials in osteoarthritis. Osteoarthritis and Cartilage. 2015;**23**(5):686-697. DOI: 10.1016/j.joca.2015.03.002

[88] Benders KE, van Weeren PR, Badylak SF, Saris DB, Dhert WJ, Malda J. Extracellular matrix scaffolds for cartilage and bone regeneration. Trends in Biotechnology. 2013;**31**(3):169-176. DOI: 10.1016/j.tibtech.2012.12.004

[89] Butoescu N, Jordan O, Doelker E. Intra-articular drug delivery systems for the treatment of rheumatic diseases: A review of the factors influencing their performance. European Journal of Pharmaceutics and Biopharmaceutics. 2009;**73**(2):205-218. DOI: 10.1016/j.ejpb.2009.06.009

[90] Jiang N, Yang P, Liu S, Li H, Wu L, Shi X, et al. Satisfaction of patients and physicians with treatments for rheumatoid arthritis: A population-based survey in China. Patient Preference and Adherence. 2020;**14**:1037-1047. DOI: 10.2147/PPA.S232578

Chapter 2

Aging and Synovial Joint Function: Changes in Structure and Implications for Mobility

Rabia Gul, Imran Nazir, Muhammad Imran Amirzada, Faryal Jahan, Faiza Naseer and Tahir Ahmad Baig

Abstract

This chapter focuses on the impact of aging on synovial joint function, specifically the changes that occur in joint structure and their implications for mobility. These age-related changes can lead to joint degeneration, decreased joint flexibility, and increased susceptibility to injury or conditions like osteoarthritis. Furthermore, the chapter discusses the consequences of these structural changes on mobility and functional abilities in older individuals. The gradual decline in joint function due to aging can result in limitations in range of motion, joint stiffness, reduced muscle strength, and overall decreased mobility. The implications of these changes on activities of daily living and quality of life are explored, highlighting the importance of maintaining joint health and promoting active lifestyles in older adults. Additionally, the chapter touches upon potential strategies and interventions to mitigate the impact of aging on synovial joint function. It covers approaches such as exercise programs, physical therapy, nutritional considerations, and joint-specific interventions that can help optimize joint health, enhance mobility, and potentially slow down the progression of age-related joint degeneration. Understanding the structural changes that occur in synovial joints with aging and their consequences for mobility is vital for healthcare professionals, researchers, and individuals seeking to promote healthy aging and maintain joint function throughout the lifespan.

Keywords: synovial joints, aging, joint physiology, physical interventions, chronic inflammation

1. Introduction

Aging is a natural biological progression that triggers various physiological transformations, notably impacting the musculoskeletal system. Changes in the structure of synovial joints assume a pivotal role in shaping an individual's mobility and overall well-being [1]. These joints, essential for facilitating movement and offering mechanical support, undergo a sequence of modifications during aging.

Such alterations entail intricate interactions among structural elements like cartilage, synovial fluid, ligaments, and the joint capsule, resulting in decreased joint function [2].

Grasping the complex association between aging and synovial joint function is crucial for comprehending the mechanisms underlying age-related constraints on mobility and disorders associated with joints. This chapter explores the multifaceted dimensions of the aging process and its consequences for the structure and function of synovial joints [3]. By meticulously examining the specific structural changes that transpire within these joints, the aim is to uncover the repercussions of these modifications on an individual's mobility and ability to carry out daily activities independently [4–6]. Additionally, the chapter will assess the implications of these age-related changes in synovial joints for developing preventive and therapeutic interventions, with the explicit objective of enhancing the quality of life for the aging population. Employing an extensive analysis of relevant literature and recent advancements in the field, the ultimate goal is to provide a comprehensive understanding of the intricate interplay between aging, synovial joint structure, and mobility, fostering enhanced management and care for individuals experiencing age-related deterioration in joint function [1, 3].

2. The anatomy of synovial joints: a foundation for understanding age-related modifications

The anatomy of synovial joints constitutes a foundational understanding crucial for comprehending the modifications that occur with aging and their subsequent impact on joint function. Synovial joints are the most common type of joint in the human body, characterized by their capacity for a wide range of motion. They are essential for facilitating various movements, such as flexion, extension, abduction, adduction, and rotation. Understanding the intricate structures within these joints provides insights into the complex mechanisms underlying age-related changes and their effects on joint health and function [3].

2.1 Articular cartilage

The articular cartilage covers the bone ends within the joint, serving as a smooth and resilient tissue that reduces friction and facilitates smooth movement between the articulating surfaces. Composed of specialized cells called chondrocytes embedded in an extracellular matrix, articular cartilage lacks a direct blood supply and nerve innervation. With advancing age, the articular cartilage undergoes structural changes, including thinning, loss of elasticity, and increased fibrillation and fissuring. These alterations can lead to reduced shock absorption, increased susceptibility to damage, and the development of degenerative joint conditions such as osteoarthritis [4, 5].

2.2 Synovial fluid

Synovial fluid is a viscous, transparent fluid that fills the synovial cavity within the joint, providing lubrication and nourishment to the articular cartilage. It also serves to reduce friction during joint movements. With age, the composition and viscosity of synovial fluid may change, leading to alterations in its lubricating properties. These changes can contribute to increased joint surface friction, discomfort, stiffness, and decreased range of motion [6].

2.3 Ligaments and tendons

Ligaments and tendons are crucial connective tissues that stabilize and support synovial joints. Ligaments connect bone to bone, while tendons connect muscle to bone. These structures help to limit excessive joint movement and prevent dislocation. With aging, ligaments and tendons may experience a decrease in tensile strength and elasticity, leading to decreased joint stability and an increased risk of injury [7, 8].

2.4 Joint capsule

The joint capsule is a fibrous, elastic structure that encloses the synovial joint, maintaining its structural integrity and containing the synovial fluid. The capsule comprises an outer fibrous layer and an inner synovial membrane. With age, the flexibility and elasticity of the joint capsule may diminish, leading to a decrease in the joint's range of motion and flexibility [5].

2.5 Synovial membrane

The synovial membrane lines the inner surface of the joint capsule and is responsible for producing synovial fluid. It plays a crucial role in maintaining a healthy joint environment by supplying nutrients to the articular cartilage and facilitating the removal of metabolic waste products. During the aging process, the synovial membrane may change, leading to the production of inflammatory mediators and an increase in inflammatory responses within the joint. This can contribute to developing joint-related pathologies, such as synovitis and other inflammatory joint conditions [9].

3. Age-associated alterations in synovial fluid composition and their influence on joint performance

Synovial fluid composition changes significantly as individuals age, impacting joint performance and overall health. The composition of synovial fluid, crucial for lubrication and nutrient supply to the joint, is affected by various age-related alterations [10].

3.1 Reduction in lubricating agents

A notable change is the decline in the production of lubricating agents like hyaluronic acid, which is responsible for the viscosity of the synovial fluid. Reduced levels of hyaluronic acid contribute to decreased lubrication between joint surfaces, leading to increased friction and wear within the joint. Consequently, this can result in joint stiffness, discomfort, and restricted range of motion, significantly influencing joint performance and mobility [11].

3.2 Altered protein concentration

Age-related modifications in the concentration of proteins within synovial fluid affect its lubricating and anti-inflammatory properties. Variations in the levels of lubricin, a glycoprotein that minimizes friction between joint surfaces, can lead to diminished joint lubrication and increased vulnerability to mechanical stress. Moreover, changes in the levels of enzymes like matrix metalloproteinases and aggrecanase contribute to the breakdown of extracellular matrix components, accelerating the progression of degenerative joint diseases, such as osteoarthritis [12].

3.3 Imbalance in inflammatory mediators

The equilibrium of inflammatory mediators within synovial fluid is vital for maintaining a healthy joint environment. Age-related alterations can disrupt this balance, leading to an increase in proinflammatory cytokines and a decrease in anti-inflammatory factors. This imbalance contributes to chronic inflammation within the joint, exacerbating joint damage and impairing overall joint performance [13].

Understanding the complex shifts in synovial fluid composition linked to aging is crucial for developing strategies to maintain joint health and function in older adults. Therapeutic methods focused on addressing changes in synovial fluid, including treatments like viscosupplementation and anti-inflammatory therapies, present encouraging avenues to alleviate the negative impacts of aging on joint performance and enhance the overall well-being of individuals facing age-related declines in joint function [14].

4. Structural deteriorations in aging synovial joints and their impact on mobility and flexibility

As individuals age, synovial joints undergo several structural deteriorations, significantly impacting mobility and flexibility. Understanding these changes is crucial for addressing the challenges faced by the aging population in maintaining functional independence and high quality of life [11–13].

4.1 Cartilage degeneration

One of the primary structural deteriorations is the degeneration of articular cartilage, which experiences a loss of elasticity and thinning with age. This degeneration reduces the smooth gliding of joint surfaces, leading to increased friction and discomfort during movement. As cartilage deteriorates, joint mobility diminishes, causing stiffness and reduced flexibility, limiting an individual's range of motion [15].

4.2 Ligament and tendon stiffness

With advancing age, ligaments and tendons lose their elasticity and become stiffer, impacting joint stability and flexibility. Reduced flexibility in these connective tissues can lead to a higher risk of injuries as the joints become less adept at absorbing impact and shock. The loss of flexibility in ligaments and tendons contributes to a decreased overall range of motion, affecting mobility and making movements more challenging for older individuals [16].

4.3 Changes in joint capsule elasticity

The joint capsule, a fibrous structure that encloses the synovial joint, changes elasticity and flexibility due to aging. This alteration limits the joint's ability to move freely, decreasing mobility and flexibility. The reduced elasticity of the joint capsule can result in joint stiffness and discomfort, further restricting the range of motion and making routine activities more challenging for older people [9, 11].

4.4 Synovial membrane inflammation

The synovial membrane, responsible for producing synovial fluid and maintaining a healthy joint environment, is susceptible to inflammation with age. Chronic inflammation of the synovial membrane can lead to increased production of inflammatory mediators, causing further deterioration of the joint structure. Inflammation in the synovial membrane can result in pain, swelling, and reduced mobility, significantly impacting an individual's flexibility and ability to perform daily activities [17].

5. Inflammatory responses and their role in accelerating age-related decline in synovial joint function

Inflammatory responses are critical in accelerating age-related decline in synovial joint function. As individuals age, the balance of inflammatory processes within the synovial joints can be disrupted, leading to chronic inflammation and various joint-related disorders. Understanding the impact of inflammatory responses is crucial for comprehending the underlying mechanisms of age-related joint degeneration and developing effective strategies to manage and alleviate the associated symptoms [18].

5.1 The role of chronic inflammation

Chronic low-grade inflammation, often called "inflamm-aging, "is a hallmark of the aging process and is closely linked to the degeneration of synovial joints. Persistent inflammation within the synovial membrane can lead to the release of proinflammatory cytokines and enzymes, which contribute to the breakdown of cartilage and other joint tissues. This chronic inflammatory state can perpetuate a cycle of tissue damage and repair, ultimately accelerating the deterioration of the joint structure [11, 19].

5.2 Impact on synovial fluid composition

Inflammatory responses can also influence the composition of synovial fluid. Elevated levels of inflammatory mediators within the synovial fluid can disrupt the balance of lubricating agents and enzymes, leading to decreased viscosity and impaired lubrication. This alteration in synovial fluid composition can increase joint surface friction, leading to pain, stiffness, and reduced mobility [19].

5.3 Effects on joint capsule and ligaments

Inflammation can affect the integrity of the joint capsule and surrounding ligaments, leading to a loss of elasticity and increased stiffness. Decreased flexibility and range of motion can significantly impact joint function and mobility. Moreover, chronic inflammation can promote the formation of adhesions within the joint capsule, further restricting joint movement and exacerbating the overall decline in joint function [20].

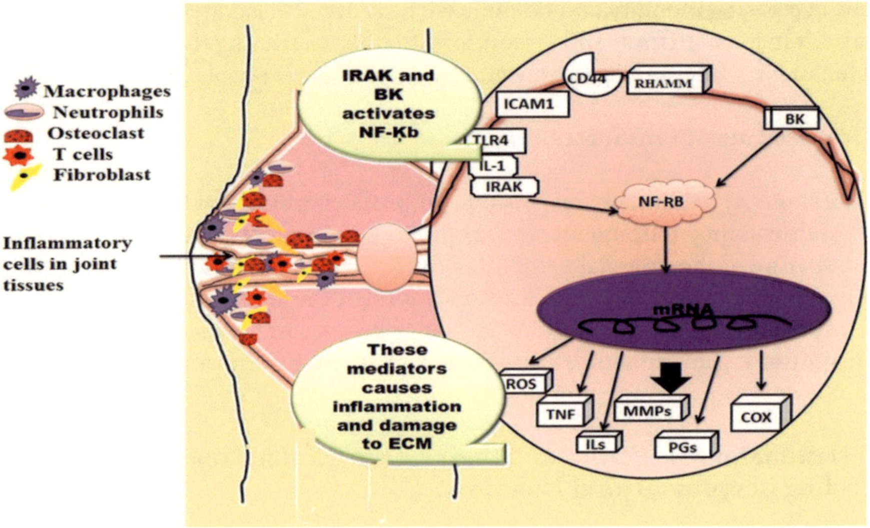

Figure 1.
Role of inflammation in degenerating synovial joints.

5.4 Role in the development of degenerative joint diseases

Chronic inflammatory responses within synovial joints are closely associated with the development and progression of degenerative joint diseases, such as osteoarthritis and rheumatoid arthritis. Inflammatory mediators can trigger the release of enzymes that degrade cartilage and bone, leading to the erosion of joint surfaces and the development of pain, swelling, and deformity. The resulting joint damage can severely compromise joint function and significantly limit an individual's mobility and overall quality of life (**Figure 1**) [17–20].

6. Implications for mobility and strategies for maintaining joint health in the aging population

The implications for mobility and strategies for maintaining joint health in the aging population are paramount, considering the significant impact of age-related changes on musculoskeletal function. Maintaining mobility is crucial for preserving independence, quality of life, and overall well-being in older individuals. Understanding the implications for mobility and implementing effective strategies for supporting joint health can significantly enhance the overall functional capacity and quality of life in the aging population [21].

As individuals age, various factors, including structural changes in joints, reduced muscle mass, decreased bone density, and alterations in balance and coordination, can collectively contribute to a decline in mobility. Age-related joint degeneration, such as cartilage thinning, ligament stiffness, and inflammation, can result in joint pain, stiffness, and reduced range of motion, significantly impacting mobility. Additionally, decreased muscle strength and flexibility can further exacerbate mobility challenges, leading to daily living difficulties and increasing the risk of falls and injuries [22].

Several strategies can be employed to maintain joint health and improve mobility in the aging population [23]. These strategies include:

6.1 Regular physical activity

Regular physical activity, including low-impact walking, swimming, and cycling, can help improve joint flexibility, muscle strength, and overall mobility. Exercise programs focused on improving balance and flexibility, such as yoga and tai chi, can also be beneficial for maintaining joint health [11, 13, 20]. Engaging in strength-training exercises helps build muscle mass and strength, providing better support and stability for the joints. Strengthening the muscles around the joints, including the quadriceps, hamstrings, and core muscles, can help alleviate the pressure on the joints, reducing the risk of joint injuries and enhancing overall joint function [21]. Physical activity stimulates the production of synovial fluid, which acts as a lubricant for the joints, promoting smoother and more efficient movement. Regular movement helps nourish the cartilage by facilitating the exchange of nutrients and waste products, thereby enabling the overall health and longevity of the joint cartilage [22]. When combined with a balanced diet, regular physical activity can aid in weight management, reducing the strain on weight-bearing joints such as the hips, knees, and ankles. Maintaining a healthy weight helps minimize the risk of developing joint-related conditions such as osteoarthritis, as excess body weight can exacerbate joint pain and accelerate the deterioration of joint structures [23]. Physical activity improves blood circulation, delivering essential nutrients and oxygen to the joints and promoting overall health and function. Improved circulation also aids in the removal of waste products and inflammatory mediators from the joints, reducing the risk of inflammation and supporting the body's natural healing processes [24]. Regular physical activity improves mood, reduces stress levels, and enhances overall well-being. The psychological benefits of exercise can contribute to better pain management, increased motivation to adhere to an active lifestyle, and improved quality of life, promoting a holistic approach to joint health and overall wellness [2, 9].

6.2 Healthy weight management

Maintaining a healthy weight is crucial for reducing the strain on joints, particularly weight-bearing joints such as the knees and hips. Excess body weight can accelerate joint degeneration and increase the risk of developing osteoarthritis, a condition characterized by the breakdown of joint cartilage [25]. By managing body weight within a healthy range, individuals can minimize the strain on their joints, reducing the likelihood of joint pain and improving overall joint function. Weight management strategies, including a balanced diet and portion control, can help alleviate joint pressure and minimize the risk of joint-related complications [26]. Maintaining a healthy weight reduces the strain on the joints and improves overall physical function and mobility. Individuals with a healthy body weight often experience improved flexibility, agility, and balance, enabling them to engage in various physical activities and daily tasks without experiencing excessive joint discomfort or limitations [27].

Healthy weight management encourages individuals to adopt a balanced and nutritious diet rich in essential vitamins, minerals, and antioxidants. Consuming a diet that includes a variety of fruits, vegetables, whole grains, lean proteins, and healthy fats can support joint health by providing the necessary nutrients for maintaining the integrity of joint structures and promoting overall musculoskeletal well-being. Healthy weight management contributes to an improved quality of life, as it helps individuals maintain their independence, engage in various activities, and experience fewer limitations related to joint health. By promoting a healthy weight, individuals can enjoy a better overall sense of well-being, reduced pain, and improved mobility, leading to a more active and fulfilling lifestyle [16, 19].

6.3 Essential nutrients for joint health

A well-balanced diet rich in essential nutrients, including vitamins, minerals, and antioxidants, can support joint health and reduce the risk of inflammation and oxidative stress. Including foods with anti-inflammatory properties, such as fatty fish, fruits, vegetables, and whole grains, can help mitigate the impact of inflammation on joint function [12].

6.3.1 Omega-3 fatty acids

Found in fatty fish such as salmon, mackerel, and sardines, as well as in flaxseeds and chia seeds, omega-3 fatty acids possess potent anti-inflammatory properties. These nutrients can help reduce joint stiffness and alleviate symptoms associated with inflammatory joint conditions such as rheumatoid arthritis [10].

6.3.2 Vitamin D

Essential for maintaining bone health, vitamin D is crucial in calcium absorption and mineralization. Sunlight exposure, fortified dairy products, and certain fish species are excellent sources of vitamin D, which can help prevent the development of osteoporosis and support overall joint integrity [14].

6.3.3 Calcium and magnesium

Crucial minerals for bone health. Calcium and magnesium help maintain bone density and strength, reducing the risk of fractures and osteoporosis. Dairy products, leafy green vegetables, nuts, and seeds are excellent sources of these minerals, supporting overall bone and joint health [7, 9].

6.3.4 Antioxidants

Abundant in fruits and vegetables, antioxidants such as vitamins C and E help neutralize free radicals, reducing oxidative stress and inflammation. Various colorful fruits and vegetables can provide a rich source of antioxidants, promoting joint health and protecting against joint-related conditions [16, 21].

6.3.5 Anti-inflammatory foods

Certain foods possess anti-inflammatory properties that help alleviate joint pain and reduce inflammation. Incorporating foods such as berries, cherries, turmeric, ginger, and green tea into the diet can help mitigate the symptoms of inflammatory joint disorders and promote overall joint health [20].

6.3.6 Hydration

Maintaining adequate hydration is crucial for joint health, as water helps hydrate and nourish the cartilage within the joints. Staying properly hydrated supports the smooth movement of joints and helps prevent the onset of conditions such as gout, which can be exacerbated by dehydration [6, 9].

6.3.7 Balanced diet for overall well-being

Emphasizing a balanced diet that includes a variety of whole grains, lean proteins, healthy fats, and a colorful array of fruits and vegetables is vital for promoting overall musculoskeletal well-being. A well-rounded diet provides the necessary nutrients for maintaining joint integrity, reducing the risk of inflammation and oxidative stress, and supporting overall joint health and mobility [1, 9].

6.4 Joint-friendly lifestyle modifications

Joint-friendly lifestyle modifications are essential for promoting and maintaining optimal joint health, particularly in individuals experiencing age-related changes or those with preexisting joint conditions. These modifications encompass various adjustments to daily activities and habits that help minimize stress on the joints, reduce the risk of injury, and alleviate joint pain [11].

6.4.1 Proper body mechanics

Adopting proper body mechanics during daily activities, such as lifting, carrying, and bending, can significantly reduce joint strain. Using the appropriate techniques, such as lifting with the legs rather than the back, can help prevent injuries, especially to the spine, hips, and knees. Maintaining proper posture while sitting, standing, and walking is also essential for reducing unnecessary stress on the spine and promoting optimal joint alignment [20].

6.4.2 Ergonomic adjustments

Making ergonomic adjustments in the home and workplace can help reduce joint strain and improve overall comfort. Using supportive chairs with proper back and armrests, adjusting the height of workstations to promote good posture, and using ergonomic tools and devices that reduce repetitive strain on the joints can significantly contribute to minimizing joint discomfort and promoting musculoskeletal health [7, 10].

6.4.3 Low-impact exercise

Engaging in low-impact exercises, such as swimming, cycling, and using elliptical machines, can help improve joint flexibility and muscle strength without placing excessive stress on the joints. These exercises promote cardiovascular health, enhance joint mobility, and support overall physical function, making them ideal for individuals with joint-related conditions or those looking to prevent joint injuries [11, 20].

6.4.4 Joint protection during physical activities

Implementing proper joint protection techniques during physical activities, such as wearing appropriate protective gear during sports or using joint-supporting braces or wraps, can help minimize the risk of joint injuries. Additionally, using the correct footwear with adequate arch support and cushioning can reduce the impact on the joints during walking, running, or other weight-bearing activities [21].

6.4.5 Avoidance of repetitive joint movements

Limiting repetitive joint movements and avoiding activities that place excessive strain on the joints can help prevent the development or exacerbation of joint-related conditions such as tendonitis or bursitis. Regular breaks during repetitive tasks and incorporating variety into daily activities can reduce the risk of overuse injuries and support long-term joint health [28].

6.4.6 Assistive devices and mobility aids

Using assistive devices and mobility aids, such as canes, walkers, and grab bars, can provide additional support and stability, particularly for individuals with mobility issues or those at a higher risk of falls. These devices help reduce the pressure on the joints, promote safer movement, and enhance overall functional independence [26].

6.5 Medical interventions and therapies

Seeking appropriate medical interventions, including physical therapy, pain management, and joint-preserving surgeries, can help alleviate joint pain, improve mobility, and enhance overall joint health and function [22].

6.6 Physical therapy

Physical therapy is a fundamental component of joint health management, focusing on rehabilitating and strengthening muscles, tendons, and ligaments surrounding the joints [5, 8]. Physical therapists develop customized exercise regimens and treatment plans tailored to the specific needs of individuals, aiming to improve joint flexibility, range of motion, and overall musculoskeletal function. Physical therapy can also help alleviate joint pain, reduce inflammation, and enhance mobility, making it an essential intervention for individuals recovering from joint injuries or those managing chronic joint-related conditions [12, 19].

6.7 Pharmacological interventions

Pharmacological interventions, including the use of nonsteroidal anti-inflammatory drugs (NSAIDs), corticosteroids, and disease-modifying antirheumatic drugs (DMARDs), are commonly employed to manage pain, reduce inflammation, and slow the progression of joint-related disorders such as osteoarthritis and rheumatoid arthritis [28]. These medications help alleviate symptoms, improve joint function, and enhance the overall quality of life for individuals experiencing joint discomfort and inflammation [20].

DOI: http://dx.doi.org/10.5772/intechopen.1003866

6.7.1 Joint injections

Intra-articular joint injections, such as corticosteroids and hyaluronic acid, provide targeted relief for individuals with inflammatory joint conditions. Corticosteroid injections help reduce inflammation and alleviate pain, while hyaluronic acid injections improve joint lubrication and cushioning, particularly in individuals with osteoarthritis. These injections can provide temporary or long-term relief, depending on the specific needs and conditions of the individual [29].

6.7.2 Joint-preserving surgeries

Surgical interventions, including arthroscopic procedures, joint realignment surgeries, and joint replacement surgeries, are employed to preserve joint function, alleviate pain, and improve overall joint mobility [2]. These surgeries are recommended for individuals with advanced joint degeneration or severe joint-related conditions that have not responded to conservative treatments. Joint-preservation surgeries aim to restore joint integrity, improve range of motion, and enhance the overall quality of life for individuals with debilitating joint conditions [30].

6.8 Complementary and alternative therapies

Complementary and alternative therapies, such as acupuncture, chiropractic care, and massage therapy, are often utilized in conjunction with conventional treatments to provide additional relief for individuals with joint-related discomfort. These therapies focus on promoting relaxation, reducing muscle tension, improving overall joint flexibility, enhancing the effectiveness of conventional medical interventions, and contributing to a comprehensive approach to joint health management [22–25].

7. Conclusion

In conclusion, this comprehensive exploration of aging and its impact on synovial joint function underscores the intricate interplay between physiological changes, structural modifications, and their implications for mobility and overall joint health. The aging process brings about a series of complex alterations within the synovial joints, including changes in synovial fluid composition, structural deterioration, inflammatory responses, and mobility limitations. These changes significantly contribute to the development and progression of various joint-related disorders, posing significant challenges for the aging population.

Understanding the multifaceted nature of these age-related modifications in synovial joints is paramount in developing effective strategies to preserve joint health and enhance the quality of life for older individuals. Implementing interventions such as regular physical activity, healthy weight management, proper nutrition, joint-friendly lifestyle modifications, and medical interventions and therapies can play a pivotal role in mitigating the adverse effects of aging on joint function and mobility.

Furthermore, fostering a holistic approach to joint health management that integrates preventive measures, personalized treatment plans, and comprehensive care can significantly improve functional independence and overall well-being in the aging population. By emphasizing the importance of early intervention, regular monitoring, and the implementation of evidence-based practices, healthcare professionals can empower individuals to actively participate in optimal joint health, promoting a more active, independent, and fulfilling lifestyle throughout the aging process.

Author details

Rabia Gul[1,2*], Imran Nazir[3], Muhammad Imran Amirzada[1], Faryal Jahan[2], Faiza Naseer[2,4] and Tahir Ahmad Baig[4]

1 Department of Pharmacy, COMSATS University Islamabad, Abbottabad Campus, Abbottabad, Pakistan

2 Shifa College of Pharmaceutical Sciences, Shifa Tameer-e-Millat University, Islamabad, Pakistan

3 Department of Pharmacy, COMSATS University Islamabad, Lahore Campus, Lahore, Pakistan

4 Industrial Biotechnology, Atta-ur-Rehman School of Applied Biosciences, National University of Science and Technology, Islamabad, Pakistan

*Address all correspondence to: rabia.scps@stmu.edu.pk

References

[1] Loeser RF. Age-related changes in the musculoskeletal system and the development of osteoarthritis. Clinics in Geriatric Medicine. 2010;**26**(3):371-386. DOI: 10.1016/j.cger.2010.03.002

[2] Walk RE, Moon HJ, Tang SY, Gupta MC. Contrast-enhanced microCT evaluation of degeneration following partial and full width injuries to the mouse lumbar intervertebral disc. Scientific Reports. 2022;**12**(1):15555. DOI: 10.1038/s41598-022-19487-9

[3] Jiongnan X, Shao T, Lou J, Zhang J, Xia C. Aging, cell senescence, the pathogenesis and targeted therapies of intervertebral disc degeneration. Frontiers in Pharmacology. 2023;**14**:1172920. DOI: 10.3389/fphar.2023.1172920

[4] Chaganti RK, Parimi N, Cawthon P, et al. Association of 25-hydroxyvitamin D with prevalent osteoarthritis of the hip in elderly men: The osteoporotic fractures in men study. Arthritis and Rheumatism. 2010;**62**(2):511-514

[5] He Y, Li Z, Alexander PG, Ocasio-Nieves BD, Yocum L, Lin H, et al. Pathogenesis of osteoarthritis: Risk factors, regulatory pathways in chondrocytes, and experimental models. Biology. 2020;**9**:194. DOI: 10.3390/biology9080194

[6] Ohnishi T, Iwasaki N, Sudo H. Causes of and molecular targets for the treatment of intervertebral disc degeneration: A review. Cell. 2022;**11**(3):394. DOI: 10.3390/cells11030394

[7] Janockova J, Slovinska L, Harvanova D, Spakova T, Rosocha J. New therapeutic approaches of mesenchymal stem cells-derived exosomes. Journal of Biomedical Science. 2021;**28**(1):39

[8] Sekelova T, Danisovic L, Cehakova M. Rejuvenation of senescent mesenchymal stem cells to prevent age-related changes in synovial joints. Cell Transplantation. 2023;**32**:1-17. DOI: 10.1177/09636897231200065

[9] Wen P, Zheng B, Zhang B, Ma T, Hao L, Zhang Y. The role of ageing and oxidative stress in intervertebral disc degeneration. Frontiers in Molecular Biosciences. 2022;**9**:1052878. DOI: 10.3389/fmolb.2022.1052878

[10] Colbath AC, Frisbie DD, Dow SW, Kisiday JD, Wayne McIlwraith C, Goodrich LR. Equine models for the investigation of mesenchymal stem cell therapies in orthopaedic disease. Operative Techniques in Sports Medicine. 2017;**25**(1):41-49. DOI: 10.1053/j.otsm.2016.12.007

[11] Zhang N, Zhu J, Ma Q, Zhao Y, Wang Y, Hu X, et al. Exosomes derived from human umbilical cord MSCs rejuvenate aged MSCs and enhance their functions for myocardial repair. Stem Cell Research & Therapy. 2020;**11**(1):273

[12] Tonomura H, Nagae M, Takatori R, Ishibashi H, Itsuji T, Takahashi K. The potential role of hepatocyte growth factor in degenerative disorders of the synovial joint and spine. International Journal of Molecular Sciences. 2020;**21**:8717. DOI: 10.3390/ijms21228717

[13] Abed E, Bouvard B, Martineau X, Jouzeau JY, Reboul P, Lajeunesse D. Elevated hepatocyte growth factor levels in osteoarthritis osteoblasts contribute to their altered response to bone morphogenetic protein-2 and reduced

mineralization capacity. Bone. 2015;**75**:111-119

[14] Frisch RN, Curtis KM, Aenlle KK, Howard GA. Hepatocyte growth factor and alternative splice variants—Expression, regulation and implications in osteogenesis and bone health and repair. Expert Opinion on Therapeutic Targets. 2016;**20**:1087-1098

[15] Ishibashi H, Tonomura H, Ikeda T, Nagae M, Sakata M, Fujiwara H, et al. Hepatocyte growth factor/c-met promotes proliferation, suppresses apoptosis, and improves matrix metabolism in rabbit nucleus pulposus cells *in vitro*. Journal of Orthopaedic Research. 2016;**34**:709-716

[16] Hang F, Zhao X, Shen H, Zhang C. Molecular mechanisms of cell death in intervertebral disc degeneration (review). International Journal of Molecular Medicine. 2016;**37**:1439-1448

[17] Shamji MF, Setton LA, Jarvis W, So S, Chen J, Jing L, et al. Proinflammatory cytokine expression profile in degenerated and herniated human intervertebral disc tissues. Arthritis and Rheumatism. 2010;**62**:1974-1982

[18] Zou F, Jiang J, Lu F, Ma X, Xia X, Wang L, et al. Efficacy of intradiscal hepatocyte growth factor injection for the treatment of intervertebral disc degeneration. Molecular Medicine Reports. 2013;**8**:118-122

[19] Conaghan PG et al. Impact and therapy of osteoarthritis: The arthritis care OA nation 2012 survey. Clinical Rheumatology. 2015;**34**(9):1581-1588

[20] Coryell PR, Diekman BO, Loeser RF. Mechanisms and therapeutic implications of cellular senescence in osteoarthritis. Nature Reviews Rheumatology. 2021;**17**:47-57. DOI: 10.1038/s41584-020-00533-7

[21] Allen KD et al. Osteoarthritis: Models for appropriate care across the disease continuum. Best Practice & Research. Clinical Rheumatology. 2016;**30**(3):503-535

[22] Vidal-Bralo L et al. Specific premature epigenetic aging of cartilage in osteoarthritis. Aging. 2016;**8**(9):2222-2231

[23] Fellows CR, Matta C, Zakany R, Khan IM, Mobasheri A. Adipose, bone marrow and synovial joint-derived mesenchymal stem cells for cartilage repair. Frontiers in Genetics. 2016;**7**:234346. DOI: 10.3389/fgene.2016.00213

[24] Luria A, Chu CR. Articular cartilage changes in maturing athletes: New targets for joint rejuvenation. Sports Health. 2014;**6**(1):18-30. DOI: 10.1177/1941738113514369

[25] Sekiya I, Muneta T, Horie M, Koga H. Arthroscopic transplantation of synovial stem cells improves clinical outcomes in knees with cartilage defects. Clinical Orthopaedics and Related Research. 2015;**473**:2316-2326. DOI: 10.1007/s11999-015-4324-8

[26] Spees JL, Lee RH, Gregory CA. Mechanisms of mesenchymal stem/stromal cell function. Stem Cell Research & Therapy. 2016;**7**:1-13. DOI: 10.1186/s13287-016-0363-7

[27] Steward AJ, Kelly DJ. Mechanical regulation of mesenchymal stem cell differentiation. Journal of Anatomy. 2015;**227**:717-731. DOI: 10.1111/joa.12243

[28] Trumbull A, Subramanian G, Yildirim-Ayan E. Mechanoresponsive musculoskeletal tissue differentiation of adipose-derived stem cells. Biomedical Engineering Online. 2016;**15**:43. DOI: 10.1186/s12938-016-0150-9

[29] Wyles CC, Houdek MT, Behfar A, Sierra RJ. Mesenchymal stem cell therapy for osteoarthritis: Current perspectives. Stem Cells Cloning. 2015;**8**:117-124. DOI: 10.2147/SCCAA.S68073

[30] Zhou Q, Li B, Zhao J, Pan W, Xu J, Chen S. IGF-I induces adipose derived mesenchymal cell chondrogenic differentiation in vitro and enhances chondrogenesis *in vivo*. In Vitro Cellular & Developmental Biology. Animal. 2016;**52**:356-364. DOI: 10.1007/s11626-015-9969-9

Chapter 3

Perspective Chapter: Exploring a Modified Portal in Shoulder Arthroscopy – A Surgical Technique for Rotator Cuff Injury and Acromioplasty

Lucas Alves Araújo, Wander Edney de Brito and Alessandro Rozim Zorzi

Abstract

In conventional arthroscopic repair of rotator cuff tears, a standard approach involves the use of three portals. However, the anterior portal poses potential risks, including cephalic vein injury leading to hematoma formation and the need for conversion to an open surgical technique, which can be particularly challenging when the patient is positioned laterally. The primary objective of this article is to elucidate an innovative technique employing a modified anterolateral portal, thereby eliminating the need for the conventional anterior portal. This modification significantly mitigates the risk of cephalic vein injury and preserves the integrity of the anterior deltoid muscle, offering a safer and more effective approach to rotator cuff repair.

Keywords: synovial joints, shoulder, arthroscopy, rotator cuff repair, surgery

1. Introduction

Rotator cuff injury (RCI) stands out as one of the most prevalent shoulder problems, accounting for approximately 23% of consultations with shoulder surgery specialists. This condition can manifest with painful symptoms, limitations in shoulder mobility and strength, often resulting in work absenteeism. It affects around 20% of the general population, and in individuals aged over 80, the incidence can reach a staggering 50% [1, 2].

While conservative treatment is an initial option, in cases of unsatisfactory progression, surgical treatment is the recommended alternative, performed through open or arthroscopic procedures. Among surgical techniques, arthroscopy is highlighted for providing optimized joint visualization with minimal surrounding tissue trauma [3, 4].

Traditionally, arthroscopic repair of RCI involves the creation of three classic surgical portals: anterior, posterior, and lateral, with the option to incorporate accessory portals. Creating arthroscopic portals in the shoulder may seem like a straightforward procedure, but there are risks associated with neurovascular structure injury near the joint. Anatomical studies in cadavers have shown that anterior portals pose the highest risk of injuries. Out of every six dissections conducted in cadavers exposed to anterocentral and anteroinferior portals, one resulted in cephalic vein injury, accounting for 16.6% of the total cadavers dissected with cephalic vein injuries following shoulder arthroscopy using these portals [5, 6].

The aim of this manuscript is to present an arthroscopic technique for RCI repair and acromioplasty, utilizing a variation of the lateral-anterior portal, with the intention of avoiding anterior portals and, in turn, reducing the associated risks.

2. Description of the technique

This study received approval from the Research Ethics Committee of the institution (CAAE 59538722.3.0000.5374; number 5,568,644 issued on August 8, 2022). The indications and contraindications remain consistent with those of the traditional anterior portal arthroscopic repair.

3. Patient positioning

On the surgical table following general anesthesia, the patient is placed in a lateral decubitus position on the side opposite to the one being operated. The upper limb to be treated is secured within a shoulder distraction system in a lateral decubitus, using a tubular mesh and sterile drapes, which facilitates abduction, anterior flexion, and traction during the surgical procedure. Traction is applied using weights approximately equal to 10% of the patient's body weight, typically ranging between 5 and 10 kg, providing adequate arthroscopic visualization. Proper head protection and cushions on all bony prominences are necessary to prevent pressure injuries and neuropraxias. The primary advantage of this position is the reduced risk of cerebral hypoperfusion.

4. Creation of portals

Following the preparation and placement of surgical drapes, the initial step involves marking the portals, utilizing three primary anatomical references: the scapular spine, the acromion, and the clavicle. The posterior portal is located 2 to 3 cm below and 1 to 2 cm medially from the posterior-lateral edge of the acromion. Subsequently, a new marking is made, starting from the medial curvature of the acromion, extending from the posterior part of the acromioclavicular joint, moving laterally across the acromion. The lateral-anterior (LA) portal is marked 2 cm distal and lateral to the acromion, immediately up to 1 cm anterior to the lateral marking (**Figure 1**).

Through the posterior portal, the trocar is introduced, followed by a 30° angled scope into the intra-articular glenohumeral space for diagnostic arthroscopy. The trocar is then inserted directly into the subacromial space with the scope facing the underside of the acromion. Before introducing the cannula into the LA portal, the

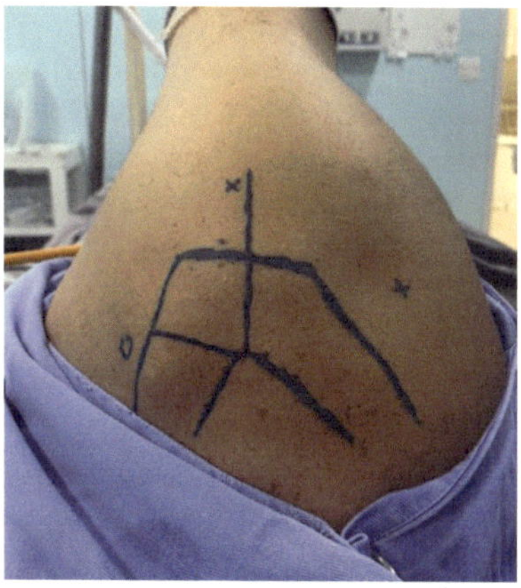

Figure 1.
Marking of the two portals for shoulder arthroscopy: The traditional posterior and the lateral-anterior (LA).

assistance of an intravenous catheter (18-gauge Jelco) is used, which can be visualized with the scope already inserted into the posterior portal.

The key differentiators of this technique are the omission of anterior portals and, notably, the accessory portal created directly lateral to the acromial edge and anterior to the lateral marking with the aid of the 18-gauge Jelco catheter for locating the most suitable anchor insertion site. A small incision of 2 to 3 millimeters is sufficient for this portal, eliminating the need for cannula insertion. Through this portal, the anchor guide is introduced, and subsequently, the anchor is secured in the optimal position for RCI repair. By rotating the limb internally and externally, the ideal anchor or anchors' positions can be determined.

5. Diagnostic arthroscopy and acromioplasty

We begin by inserting a 4.0 mm, 30° angled optic scope coupled to a camera with a light source through the posterior portal into the subacromial space. After visualizing the underside of the acromion, the scope is directed toward the lateral region of the shoulder. Using an 18-gauge intravenous catheter (Jelco 18) inserted into the previously marked LA portal, the optimal position for cannula insertion is confirmed.

Once the cannula is introduced through the LA portal, other instruments are sequentially inserted through this portal. Soft tissue debridement is performed using a shaver blade and radiofrequency probe to remove devitalized tissues and expose the undersurface of the acromion. Subsequently, acromioplasty and lateral clavicle resection are carried out with the bone shaver, if required. The scope is then directed to the patient's humerus, allowing visualization of the subacromial bursa, which is then excised during bursectomy.

At this point, the rotator cuff is visible. Under direct visualization through the scope, the extent of the lesion is assessed, and the most suitable anchor placement location is determined. Irregular edges of the damaged rotator cuff are debrided using the soft tissue shaver, and the humeral surface, which will serve as the footprint for rotator cuff reattachment, is decorticated with the bone shaver. Once again, with the aid of the intravenous catheter (Jelco 18) inserted immediately lateral to the acromion, the optimal anchor insertion site is confirmed to ensure that it is positioned at approximately a 45° angle relative to the humeral surface. A small incision of 2 to 3 millimeters is made at the location where the venous catheter was inserted to create the accessory portal.

Through this accessory portal, the anchor guide is introduced, and subsequently, the anchor is secured in the most appropriate position for rotator cuff repair. By externally and internally rotating the operated limb, the optimal anchor position can be identified. The anchor, loaded with two suture threads, is inserted at a 45° angle relative to the humerus. Each suture thread is separately secured (**Figure 2**).

One of the suture threads, in the most suitable position, is selected, and its two ends are separated. One end is brought through the LA portal and fastened to a suture passer with a retrieval system. The suture passer is introduced through the cannula and is used to repair the rotator cuff. The end of the suture thread is brought back through the LA portal and secured. The other end of the same suture thread is also brought to the LA portal. With both ends of the thread through the LA portal, the self-locking "SMC" knot is used for the repair and fixation of the cuff to the anchor and bone.

If necessary, the same procedure can be performed with the other suture thread loaded in the anchor already inserted. Additionally, more anchors can be used if the extent of the lesion requires it, using the same accessory portal for anchor insertion

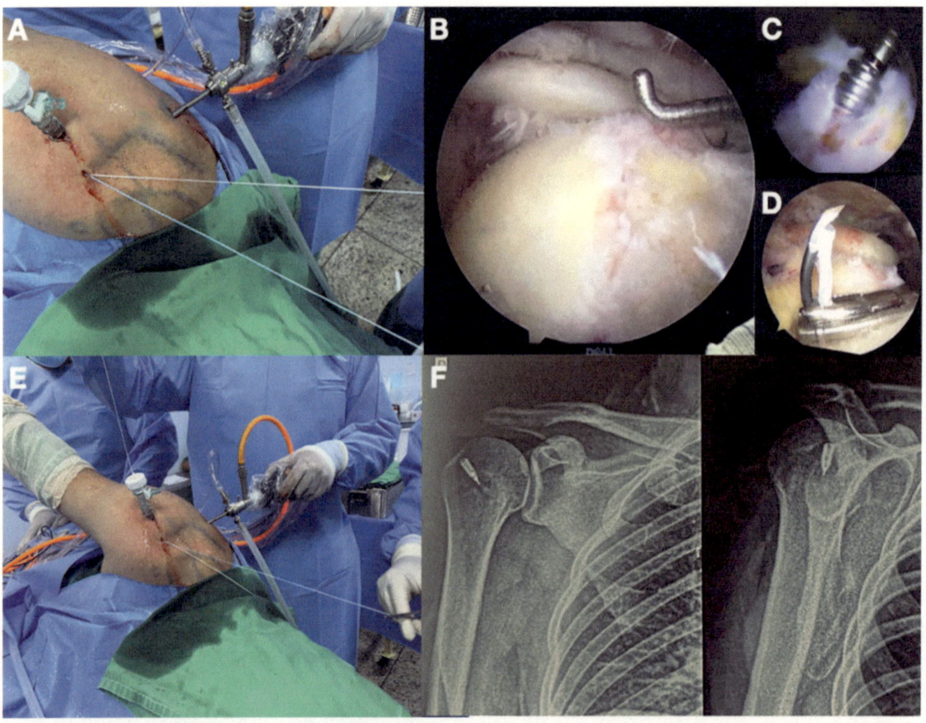

Figure 2.
Repair of the rotator cuff injury using the two-portal technique; A) patient in lateral decubitus position, with the scope introduced through the posterior portal and the cannula in the lateral-anterior (LA) portal. An accessory incision is used to insert the anchor; B) arthroscopic image of the shoulder with a rotator cuff injury; C) insertion of the anchor into the humeral head; D) passage of the suture through the cuff; E) retrieval of the thread and knot through the LA portal; F) shoulder X-rays with a metallic anchor in the humeral head.

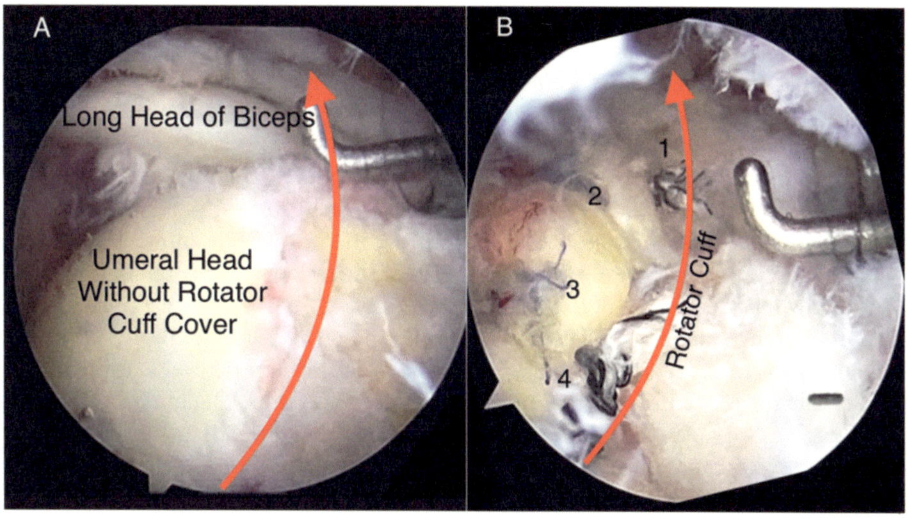

Figure 3.
A) Extensive rotator cuff injury; B) repaired with two anchors and four sutures using the two-portal and accessory technique.

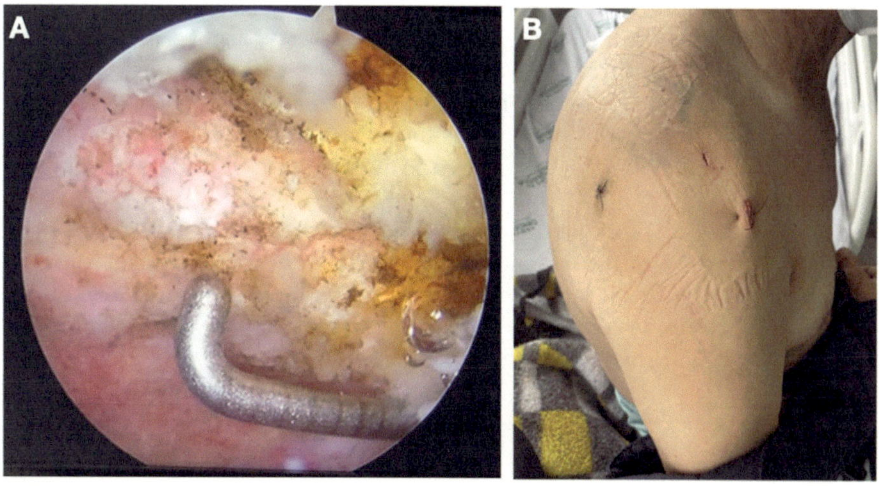

Figure 4.
A) Acromioplasty through the LA portal; B) final appearance of the surgical scar with two portals and an accessory.

and rotating the limb to locate the optimal position. This allows for the technique to be employed even for extensive lesions (**Figure 3**). If required, acromioplasty can also be performed through the LA portal (**Figure 4**).

6. Final comments

The primary outcome of this study is the detailed description of the surgical technique for repairing rotator cuff injuries (RCI) and performing acromioplasty using a modified lateral-anterior portal (LA) while avoiding the use of the anterior portal. This technique offers several key advantages, including a reduced risk of cephalic vein injury, less trauma to the deltoid muscle (particularly its anterior portion), and cost savings due to the use of a single cannula.

Other authors have previously proposed various portal approaches for shoulder RCI repair. One technique involved creating arthroscopic portals based on the location of the rotator cuff tear. However, this approach utilized four portals: posterior, anterior, lateral, and anterosuperior to the suprascapular fossa. In our opinion, the use of the anterior portal does not resolve the issue of the risk of cephalic vein injury with significant bleeding and the potential for conversion to open surgery [7].

Another technique was suggested, using only the posterior and anterolateral portals. The author introduced a variation in which the anterolateral portal is positioned more distally for repairing the long head of the biceps tendon. The most significant finding of the study was that 11 out of 23 portals (47.8%) in the distal anterolateral location were in contact with a distal branch of the axillary nerve. The proximity and frequent contact of this portal with the distal axillary nerve branches on the undersurface of the anterior deltoid necessitate caution when placing these portals. Since our technique focuses on RCI repair, the lateral-anterior portal described here is more lateral and proximal, reducing the risk of axillary nerve branch injury [8].

In another described technique, two portals are used, a single anterolateral working portal in conjunction with a posterior viewing portal, allowing for several procedures, including subacromial bursectomy, acromioplasty, distal clavicle excision, supraspinatus tendon rupture repair, and long head of the biceps tenodesis. However, this study differs from the technique proposed here because the patient is placed in a beach chair position, and the location of the anterolateral portal is more anterior compared to our approach. In our opinion, the more anterior the portal, the higher the risk of venous injury [7].

The limitation of this study is the absence of a case series with data to support the safety and effectiveness of the technique. Although the author has already successfully performed a significant number of cases, a prospective clinical study will be conducted in the future to compare the efficacy and safety of the two-portal technique to the traditional anterior portal approach.

Author details

Lucas Alves Araújo, Wander Edney de Brito and Alessandro Rozim Zorzi*
São Leopoldo Mandic Medical School, Campinas, SP, Brazil

*Address all correspondence to: arzorzi@hc.unicamp.br

References

[1] Yamamoto A, Takagishi K, Osawa T, et al. Prevalence and risk factors of a rotator cuff tear in the general population. Journal of Shoulder and Elbow Surgery. 2010;**19**(01):116-120

[2] Malavolta EA, Gracitelli MEC, Assunção JH, Pinto GMR, da Silveira AZF, Ferreira AA. Shoulder disorders in an outpatient clinic: An epidemiological study. Acta Ortopédica Brasileira. 2017;**25**(03):78-78

[3] Godinho GG, Souza JMG, Bicalho LA. Reparo das rupturas do manguito rotador do ombropela videoartroscopia cirúrgica: técnica. Revista Brasileira de Ortopedia. 1996;**31**(4):284-288

[4] Garrett WE, Swiontkowski MF, Weinstein JN, et al. American Board of Orthopaedic Surgery Practice of the orthopaedic surgeon: Part-II, certification examination case mix. The Journal of Bone and Joint Surgery. American Volume. 2006;**88**:660-667

[5] Meyer M, Graveleau N, Hardy P, Landreau P. Anatomic risks of shoulder arthroscopy portals: Anatomic cadaveric study of 12 portals. Arthroscopy. 2007;**23**(5):529-536. DOI: 10.1016/j.arthro.2006.12.022

[6] Aouad D, El Rassi G. Shoulder arthroscopic rotator cuff repair with biceps Tenodesis and Acromioplasty using a single working portal. Arthroscopy Techniques. 2021;**10**(4):e1125-e1129. DOI: 10.1016/j.eats.2021.01.005

[7] Kim SH, Ha KI, Ahn JH, Park JH. Differential arthroscopic portal placement for rotator cuff repair. Arthroscopy. 2002;**18**(8):E43. DOI: 10.1053/jars.2002.29941

[8] Knudsen ML, Hibbard JC, Nuckley DJ, Braman JP. The low-anterolateral portal for arthroscopic biceps tenodesis: Description of technique and cadaveric study. Knee Surgery, Sports Traumatology, Arthroscopy. 2014;**22**(2):462-466. DOI: 10.1007/s00167-013-2444-9. Epub 2013 Feb 12

Chapter 4

Pathologies That Can Lead to Total Hip Arthroplasty

Guadalupe Irazú Morales-Reyes, Jessica Paola Plascencia-Roldán, Gilberto Flores-Vargas, María de Jesús Gallardo-Luna, Efraín Navarro-Olivos and Nicolás Padilla-Raygoza

Abstract

Hip replacement has evolved over the years, previously considered exclusively for geriatrics. It changed until it became an elective surgery in various pathologies. Certain conditions predominantly affect one age group. Hip dysplasia predominates in pediatrics, hip osteoarthritis in adults, and post-traumatic in geriatrics. Therefore, the indications for carrying out this procedure vary depending on age.

Keywords: developmental dysplasia of the hip, legg-calve-perthes disease, transient hip synovitis, epiphysiolysis of the femoral head, arthrosis

1. Introduction

Osteoarthritis is the chief indication for placing a total hip prosthesis (THP) [1]. It is a pathology with multiple and complex etiology, where there are alterations in the physiology of the cartilage and the chondrocyte -one crucial cell in charge of the metabolism of the extracellular matrix-. It is classified into primary, generated by anatomical alterations or joint degeneration without an apparent cause, and secondary, where joint damage is due to aging and related to obesity, diabetes, metabolic syndrome, and chronic and inflammatory diseases [2].

The risk factors for developing secondary osteoarthritis are hip dysplasia, femoroacetabular impingement, avascular necrosis, juvenile arthritis, septic arthritis, slipped capital femoral epiphyseal, hip or acetabular fractures, and Perthes disease [3].

Surgical treatments are advisable under two clinical contexts. In those patients with unicompartmental osteoarthritis and alteration in the axis -in which a surgical intervention (osteotomy or unicompartmental prosthesis) can improve the symptoms and anatomical alteration- in such a way that they manage to reduce the progression to the generalized joint degenerative phenomenon. The other group with surgical indication is those who did not show improvement with conservative treatment, mainly due to pain progression or decreased functionality and loss of range of motion [4].

Rheumatic pathologies are considered an indication to place a total hip prosthesis, mainly affecting the knee and hip. The reported incidence varies from 65–90% for the former and 15–36% for the latter [1, 5, 6].

Patients who develop coxarthrosis caused by rheumatoid arthritis (RA) present with painful hips, decreased mobility, flexion deformity, and external rotation. Approximately 15% will need a total arthroplasty of one or both hips -the gold standard- [5, 6].

Total hip replacement (THR) is a procedure that significantly improves the quality of life of patients with RAwith severe and limiting involvement of their hips [7].

Total hip replacement has evolved, especially in the last 50 years. Previously, it was considered an exclusive procedure for geriatrics with a low expectation of recovery. However, this has evolved and has become the surgery of choice for various hip pathologies [1].

It is considered a dynamic and evolving surgical procedure. Thanks to modern technology and instrumentation, they make this reconstructive procedure, especially in severely disabled patients, is highly predictable and cost-effective [8].

Total hip arthroplasty (THA) is considered one of the most successful reconstructive surgical procedures, consistently rated as an excellent cost-effective surgery [8].

Charnley et al. [9] proposed to treat THA and surgical access through osteotomy of the greater trochanter. Over time, surgical accesses that did not require it were preferred, thus avoiding complications such as nonunion, broken wires, and prolonged operating times [8].

Generally, revision surgery may be necessary mainly during the first or second year (due to infection, dislocation, or periprosthetic fracture) and maintained subsequently with a rate of less than 1% annually. After 20 years, revision rates increase because of osteolysis and aseptic loosening. The high survival rates are supposed to continue for 25–30 years [8].

2. Pediatrics

Regarding pediatric patients, we find diseases associated with developmental disorders such as developmental dysplasia of the hip (DDH), Legg-Calve-Perthes disease, and transient hip synovitis -which has an unknown etiology- [1].

2.1 Developmental dysplasia of the hip

Developmental dysplasia of the hip is considered a condition that can cause significant disability if not correctly treated [10].

The term encompasses different conditions, ranging from instability, dislocation, or subluxation to abnormalities detected by imaging studies such as radiography [10].

There are different types of dysplasia depending on the etiology. Its classification goes from teratological hip luxation, associated with an underlying disease and manifesting prenatally, to typical developmental dysplasia, which can be congenital or have an onset in patients without underlying disease [10].

A total of 90% of patients who do not receive timely detection, and therefore proper treatment, have moderate to severe joint disease in adulthood, requiring hip replacement at an early age, harming their social environment [10].

2.2 Legg-calve-perthes disease

Legg-Calve-Perthes disease is an orthopedic pathology affecting hip development in infants because the irrigation of the capital femoral epiphysis presents ischemic alterations, culminating in necrosis [11].

It consists of necrosis of bone tissue, articular cartilage with the proliferation of chondrocytes in the superficial layers, and dead tissue in the deep layer with the possibility of subchondral fracture of the adjacent bone and a rupture of the growth plate, progressing to bone resorption and re-ossification [11].

All of this is reflected in an X-ray study, which presents with the femoral head in the shape of a mushroom, shortness, and neck thickness; acetabulum reshaping, tilted with the shortened shaft; and metaphyseal rarefaction [11].

The phenotype is described as the elevation of the greater trochanter of the femur and coxa vara [11].

2.3 Septic arthritis and osteomyelitis

Osteoarticular infections (OAI) are more frequent in children under 5 years of age, with septic arthritis (SA) presenting at a lower age than in patients with acute osteomyelitis (OM) [12].

The most frequently affected joints are the knee (56%) and hip (26%), presenting the most common OM in the bones of the foot, femur, and tibia [12].

Staphylococcus aureus is considered the causative agent of this pathology. It produces acute infections that usually present with pain, inflammatory changes on the affected joint, fever, leukocytosis, and elevation of acute phase reactants (APR) [12].

It usually produces fever, dysregulated general state, localized skeletal pain, and functional impairment in pediatric patients. It is worth noting that not presenting fever does not discard the disease. *Kingella kingae* has been isolated, especially in children under three years of age, and it may not produce fever, leukocytosis, and elevated reactants [12].

2.4 Transient hip synovitis (THS)

It is the most frequent cause of inflammatory arthritis in pediatrics. Its etiology is unknown. It most frequently affects males between 3 and 10 years and is self-limited. It usually presents in children with no significant history, fever, or general condition, such as a sudden onset of the limp, accompanied by pain in the groin or, in 20–30% of patients, in the middle third of the thigh or knee. In 5% of cases, synovitis is bilateral [13].

The examination is characteristic and very important, with painful limitation of the last flexion degrees and, especially, internal rotation of the affected hip. The pain improves with the rest. Mobility is less painful after 48–72 hours [13].

2.5 Epiphysiolysis of the femoral head

It is an inferior and posterior slippage of the proximal femoral epiphysis on the femoral metaphysis of unknown etiology. In 20–50% of patients, it can be bilateral. It characteristically affects adolescents (mean age 12–13 years) who are obese, more frequently males (1.5:1) [13].

The most recent series, however, detect both an increase in its incidence and a decrease in the age of the patients, attributable to the growing epidemic of childhood obesity [13].

Presentation is limping or pain in the groin, thigh, or knee; frequently, patients have presented pain or discomfort in a specific location for months before they become continuous or limiting. On examination, the internal rotation blockage of the affected hip is very characteristic, and – when the hip passively flexes to 90° with the patient supine- the hip deviates into external rotation (Drehmann's sign) [13].

The confirmation is through an X-ray study of the hips. It is advisable to order an axial view as slippage of the femoral head is more evident on this view than on the AP view. Epiphysiolysis of the femoral head is an orthopedic emergency, so these patients should be immediately referred to traumatology [13].

Treatment for stable epiphysiolysis is *in situ* fixation with screws or open reduction with safe dislocation of the hip in moderate/severe displacement [14]. Unstable epiphysiolysis requires surgical treatment with gentle closed reduction and screw fixation [15].

Avascular necrosis is a complication that affects 50% of patients with unstable epiphysiolysis and deforms the femoral head causing pain, joint stiffness, osteoarthritis, and, in the long term, results in THA [16].

3. Teenagers

3.1 Juvenile idiopathic arthritis

It is defined as an inflammatory arthropathy of autoimmune etiology that presents a chronic course of at least six weeks and is considered the leading cause of chronic arthritis in childhood. The onset of the disease, the presence of rheumatoid factors, and the course of the disease are considered fundamental in the prognosis [17].

It is more affected by the female sex. According to its onset form, it can be classified as oligoarticular, which is the most frequent, polyarticular, and systemic [18].

Chronic synovitis causes antalgic postures that promote the development of flexion deformities. It is accompanied by synovial proliferation and invasive pannus deformation producing joint destruction, pericapsular adhesions, local growth alterations, demineralization, and destruction of adjacent bones. These changes generate a muscular imbalance that results in a biomechanical alteration with progressive functional loss and deformities that can progress to bone ankylosis [18].

The diagnosis is fundamentally clinical, and there is no paraclinical examination to rule out or confirm the disease. Synovial fluid analysis indicates a sterile inflammatory process, elevated protein levels, glucose at normal levels, and poor mucin clot formation. The radiological changes at the beginning of the disease are absent, so they are not helpful for early diagnosis [18].

The treatment aims to alleviate joint pain and suppress the inflammatory process. Pharmacological treatment and physiotherapy are given as the first option [17]. However, the presence of traumatic injuries generating damage to the joints is one of the reasons that hip replacement surgery is necessary. Other reasons are deteriorating or destroying joints, having a slow growth rate, uneven growth of a limb, either a leg or an arm, loss of vision, or decreased visual acuity caused by chronic uveitis and pericarditis [19].

4. Adults

In young patients, systemic diseases lead to poor bone quality, acetabular and femoral defects from congenital conditions, post-traumatic osteoarthritis, post-surgical stiffness, osteonecrosis, severe deformities, length discrepancy, pelvic obliquity, and obesity [3].

Hip arthrodesis was previously the surgical treatment of choice since it improved pain. Nevertheless, the patient had limited function. With THA, the aim is to improve the patient's functionality and quality of life. They have fewer complications. Also, mortality has decreased. Most causes of mortality are secondary to heart disease and thromboembolic pathologies [3].

One of the problems with the hip replacement technique in people aged 20–25 compared to people aged 50 lies mainly in the skeletal maturation time, since it continues after physeal closure, in addition to the fact that bone remodeling occurs faster in younger patients and influences bone-implant interactions, and they have less discipline to control the physical activity they undergo after the period of disability, leading to the need for intervention [20].

It is worth considering that Total Hip Replacement (THR) generates irreversible bone loss, and a surgical revision will probably be necessary when the patient is still young. Therefore, arthroplasty is considered very effective in the short and long term. In the medium term, it should not be advised as the first-line procedure in very young patients [20].

4.1 Arthrosis

Osteoarthritis (OA) is the most common joint disorder worldwide. Approximately 18% of women and 10% of men above 60 years old have symptomatic OA. More than half of people around 65 years old have radiological evidence of OA [21].

Hip osteoarthritis has different causes depending on the age of the patient. In people over 65 years, primary osteoarthritis occurs more frequently, which is of a degenerative type and leads the patient to present progressive joint pain that produces functional limitation to a variable degree. In patients under 40, alterations secondary to complex traumas or severe degrees of childhood pathologies that lead to corrective procedures and later joint degeneration are more frequently found [3].

The United States population suffers from an epidemic of obesity and a prevalence of primary osteoarthritis that has increased significantly. Surgical treatment with THR is recommended when there is a failure in conservative treatment, such as weight loss, increased physical activity, use of a cane, and when the administration of non-steroidal anti-inflammatory drugs fails to relieve pain [1].

The number of obese patients requiring THA has increased, so orthopedic surgeons performing this procedure must be aware of the potential problems and reduce the complications associated with these patients [22].

Secondary osteoarthritis implies greater complexity for the surgeon since patients have higher bone deformities, joint stiffness, shortening, previous surgical procedures, and instability due to muscle, capsular, or bone deficiency [3].

Depending on the stage of the disease, a variety of non-surgical and surgical treatment options are available for the management of hip osteoarthritis. Patient education, exercise therapy, and maintaining physical activity are important during the initial stages of the disease. For mild to moderate clinical data, it is possible to delay THR surgery for a while by combining these two therapies. When symptoms deteriorate in the advanced stage, THR is a successful and effective treatment [23].

Femoral fractures and osteoarthritis of the hip (OA of the hip) are the most common disease of the hip joint, treated by hip replacement surgery [23].

4.2 Post-traumatic arthritis

Post-traumatic arthritis can develop years after an acetabular fracture, impairing joint function and resulting in significant chronic musculoskeletal pain. This study systematically reviewed the literature on THA results in patients with THA and previous acetabular fracture [24].

Post-traumatic osteoarthritis, secondary to fractures and or dislocation of the acetabulum and proximal femur, is a THP indication [1].

Despite the difficulties associated with performing THA in patients with PTA from previous acetabular fracture (including soft tissue scarring, existing hardware, and acetabular bone loss) and the relatively high complication rates, THA in patients with PTA following prior acetabular fracture leads to significant improvement in pain and function at 10-year follow-up. Further high-quality randomized controlled studies are needed to confirm the outcomes after delayed THA in these patients [24].

4.3 Rheumatoid arthritis

Rheumatoid arthritis (RA) is a chronic inflammatory disease characterized by polyarthritis with progressive joint wear, increasing physical limitation, and immunological abnormalities.

It is considered an incurable condition of unknown origin, which will have to receive treatment. However, it will not be cured [25].

It affects the female sex in a ratio of 3:1. It is highly prevalent, affecting 1.6% of the Mexican population, which leads to a high economic burden for the public health system and impacts patients' quality of life [25].

In coxarthrosis secondary to RA, approximately 1 in 6 patients will need a THA [7].

Hip replacement is considered a surgical option for patients with terminal joint destruction secondary to osteoarthritis and RA [25].

Surgical correction is effective for patients with advanced joint destruction [25].

The main goal is to achieve a mobile joint with a pain-free range of motion and mechanical stability [25].

Postoperative surgical complications in replacement and their relationship with risk factors, such as a higher body mass index (BMI), preoperative use of corticosteroids, and a low serum albumin level, increase hospital readmission [25].

The problem of major surgical procedures like hip replacement is not complication-free. Recent studies show an increase in the complication proportion in the population [25].

4.4 Avascular necrosis

Osteonecrosis with the segmental collapse of the femoral head has increased in the United States due to HIV-positive patients receiving highly active antiretroviral therapy, the high rate of alcoholism in the general population, and the use of corticosteroids for various conditions. Therefore, osteonecrosis is a frequent indication for performing a THR [1].

It is considered a multifactorial pathology characterized by the progressive destruction of the bone in the coxofemoral joint due to the alteration of local blood flow. For bone tissue necrosis to develop, ischemic events must be constant [26]. Pain is considered the reason for consultation in most cases, which occurs in the groin, followed by pain in the thighs and buttocks. It is exacerbated or produced by movement and weight bearing [27]. Its incidence increases in adults. Over 75% of patients present it between 30 and 60 years, and the average age of presentation is 36 years [28]. It is most frequent in men [29]. It is associated with trauma such as femur fractures, prolonged use of glucocorticoids, and excessive alcohol consumption.

Treatment aims to preserve the hip for as long as possible. There are two approaches for osteonecrosis: conservative and surgical management [30].

Conservative management is proposed only in early stages that present small lesions or in cases of contraindicated surgical treatment. Weight-bearing restriction, pharmacological treatment, and biophysical modalities are included [31]. Surgical treatment prior to collapse involves procedures such as central decompression and non-vascularized and vascularized bone grafts. In advanced stages with a collapse and arthritic hip, hip prosthetic surgery is considered [32].

Prosthetic hip replacement has a life span of 15 years before it wears out and requires revision. For this reason, it is an optimal option for older patients [33].

Sickle cell anemia is a pathology associated with avascular necrosis of the femoral head. Deformed red blood cells cause vascular congestion, venostasis, and thrombosis of the bone microvasculature. Ischemia is aggravated by increased intraosseous pressure secondary to marrow hyperplasia, producing bone infarcts and necrosis [34]. Non-surgical treatment consists of red blood cell transfusion therapy for preventing manifestations of sickle cell anemia. THA with uncemented components is recommended in patients with sickle cell anemia as this material helps to prevent complications such as aseptic loosening by methyl methacrylate [35].

4.5 Paget's disease

Paget's disease of bone is a metabolic bone disease. Its etiology is unclear. It starts with increased resorption followed by a phase of aberrant osteoformation [36].

Approximately 10% of patients with Paget's disease develop hip osteoarthritis due to the disease itself, mainly to the alterations in the load axes that condition this pathology. Even as the establishment of early treatments decreases the incidence, symptomatic arthritis of the hip continues to be a disabling problem [37].

Hip arthroplasty is the most effective treatment for hip osteoarthritis in a patient with Paget's. However, in carrying out arthroplasty, multiple complications appear in these patients due to the specific characteristics of their bones [37].

The first difficulty orthopedic surgeons face while implanting a prosthesis is the bone deformity usually generated by the disease. It complicates arthroplasty with standard components or its positioning in varus [37].

5. Geriatrics

Falls are found more frequently in older adults. Approximately 50% occur in institutions, and 30% occur in the community [38].

A total of 90% of hip fractures are due to a fall, but only 14% of a fall results in a fracture [38].

Among the risk factors are mainly age-related changes, due to aging, such as vision alterations, muscle weakness, and proprioception alterations [38].

Pathological processes also play a role, with osteoarthritis being the most common cause, neurological problems such as Parkinson's disease, and cognitive impairment such as senile dementia, depression, anxiety, and agitation [38].

The environmental factors that influence falls in the home are obstacles to walking (carpets, cables, etc.), poor room lighting, absent bathroom supports, and inappropriate footwear [39].

There is an increase in the prevalence of hip fractures secondary to falls in patients older than 65, mainly affecting women. Women after 50 have twice the risk of suffering a hip fracture. The most frequent treatments are osteosynthesis and partial or total arthroplasty [40].

5.1 Osteoporosis

It is a disease characterized by decreased bone mass, with alteration in the bone microarchitecture, so that fragility increases and, consequently, the tendency to a possible fracture [38].

The factors related to developing osteoporosis are age, diet, environment, hereditary, and hormonal [38].

After the trauma generated by a fall, the hip fracture presents mainly with localized pain in the inguinal region, sometimes presenting irradiation toward the knee [38].

Usually, femoral head fractures, which are also displaced, present intense pain and walking impairment. There is a disability to move the hips. The affected limb is in external rotation with shortening and muscle weakness [38].

There are two options for treating hip fractures: conservative or surgical. Both aim to reduce pain and recover the patient's functional capacity [38].

Conservative treatment consists of immobilizing the patient for several months, sometimes with traction, but generating complications such as pressure ulcers, thromboembolic ulcers, and urinary and respiratory infections, so, with the introduction of osteosynthesis, this treatment had less relevance [38].

Surgical treatment must be performed within the first 48 hours of the hip fracture. Osteosynthesis or partial or THR is considered [41].

THA is indicated for patients with a displaced intracapsular hip fracture, patients of 70 years or older with a femoral neck fracture, and those over 65 years without a prior reduction, coxarthrosis, and in case osteosynthesis has failed [41].

THR has been shown to have better functional results and fewer complications for displaced femoral neck fractures versus traditional internal fixation techniques or hernioplasty in patients older than 60 years [42].

This procedure allows patients early loading, lowers the risk of requiring a second intervention, prevents failed fixation and union, and avoids avascular necrosis [41].

6. Total hip arthroplasty

Let us remember that the hip is a congruent joint, where the acetabulum and the femoral head have a symmetry that allows rotation around an axis and favors the action of the muscles [43].

Prosthesis success depends on factors related to the patient, such as age, gender, height or weight, medical history, underlying hip condition, and previous surgeries [43].

Certain surgical factors also play a role, such as the experience of the orthopedic surgeon, surgical approach, prosthesis design, component orientation, limb length inequality, and trochanteric non-union [43].

Therefore, evaluating the performance of the prosthesis becomes difficult. In any case, the average lifespan is 20 years [43].

7. Conclusions

Some various pathologies and complications lead to a THR. These pathologies occur throughout the patients' lives according to the etiology.

THR is a procedure that significantly improves the patient's quality of life.

Previously, it was thought exclusively for geriatrics with a low expectation of recovery. However, this evolved until it became the surgery of choice for various hip pathologies.

At present, it is considered a dynamic and evolving surgical procedure, which according to the different age groups, some pathologies predispose to this surgery, such as the case of pediatrics with developmental dysplasia of the hip, adults with osteoarthritis of the hip, in older adults or geriatrics with post-traumatic injuries.

Conflict of interest

The authors declare no conflict of interest.

Author details

Guadalupe Irazú Morales-Reyes[1], Jessica Paola Plascencia-Roldán[1], Gilberto Flores-Vargas[1], María de Jesús Gallardo-Luna[1], Efraín Navarro-Olivos[2] and Nicolás Padilla-Raygoza[1*]

1 Department of Research and Technological Development, Directorate of Teaching and Research, Institute of Public Health from Guanajuato State, Guanajuato, México

2 Directorate of Teaching and Research, Institute of Public Health from Guanajuato State, Guanajuato, México

*Address all correspondence to: npadillar@guanajuato.gob.mx

Preoperative serum albumin as a predictor of complications following total hip replacement in patients with rheumatoid arthritis. Acta Ortopédica Mexicana. 2018;**32**(4):193-197

[26] Deleanu B, Prejbeanu R, Vermesan D, Honcea L, Mioc ML, Tsiridis E, et al. Avascular necrosis of the femoral head at 2 years after pertrochanteric fracture surgery: Case report. Annals of Medicine and Surgery (London). 2015;**22**(5):106-109. DOI: 10.1016/j.amsu.2015.12.053

[27] Paoloni J. Approach to the adult with unspecified hip pain. In: Post TW, editor. UpToDate. Waltham, MA: UpToDate Inc; 2019. Available from: https://www.uptodate.com/contents/approach-to-theadult-with-unspecified-hip-pain?search=Approach-to-the-adult-with-unspecified-hip-pain

[28] Pouya F, Kerachian MA. Avascular necrosis of the femoral head: Are any genes involved? The Archives of Bone and Joint Surgery. 2015;**3**(3):149-155. Available from: https://www.ncbi.nlm.nih.gov/pmc/articles/PMC4507066/#:~:text=Inherited%20Avascular%20Necrosis&text=Although%20genetic%20factors%20have%20been,pathway%20in%20a%20Japanese%20family

[29] Hsu H, Nallamothu SV. Hip Osteonecrosis. StatPearls. Treasure Island, FL: StatPearls Publishing; 2023. Available from: https://www.ncbi.nlm.nih.gov/books/NBK499954/

[30] Jones LC, Mont M. Osteonecrosis (avascular necrosis of the bone). In: Post TW, editor. UpToDate. Waltham, MA: UpToDate Inc; 2018. Available from: https://www.uptodate.com/contents/osteonecrosis-avascular-necrosis-ofbone?search=Osteonecrosis-(avascular-necrosis-of-bone

[31] Petek D, Hannouche D, Suva D. Osteonecrosis of the femoral head: Pathophysiology and current concepts of treatment. EFORT Open Reviews. 2019;**4**(3):85-97. DOI: 10.1302/2058-5241.4.180036

[32] Tripathy SK, Goyal T, Sen RK. Management of femoral head osteonecrosis: Current concepts. Indian Journal of Orthopaedics. 2015;**49**(1):28-45. DOI: 10.4103/0019-5413.143911

[33] Lespasio MJ, Sodhi N, Mont MA. Osteonecrosis of the hip: A primer. The Permanente Journal. 2019;**23**:18-100. DOI: 10.7812/TPP/18-100

[34] Matos MA, dos Santos Silva LL, Brito Fernandes R, Dias Malheiros C, Pinto da Silva BV. Avascular necrosis of the femoral head in sickle cell disease patients. Ortopedia, Traumatologia, Rehabilitacja. 2012;**14**(2):155-160. DOI: 10.5604/15093492.992286

[35] Moran MC, Huo MH, Garvin KL, Pellicci PM, Salvati EA. Total hip arthroplasty in sickle cell hemoglobinopathy. Clinical Orthopaedics and Related Research. 1993;**294**:140-148. Available from: https://pubmed.ncbi.nlm.nih.gov/8358906/

[36] Ospina-Caicedo AI, Gómez-Escobar VE, Coy-Urrea VA, Segura-Charry JS, Izquierdo-Loaiza JH. Enfermedad de Paget de hueso esporádica. Serie de casos y revisión de la literatura. Revista Colombiana de Reumatologica. 2020;**27**(2):103-111. Available from: https://www.elsevier.es/es-revista-revista-colombiana-reumatologia-374-articulo-enfermedad-paget-hueso-esporadica-serie-S0121812319300647

[37] Ibor I, Merino J, López-Vinagre E, Larrea A, Rodríguez-Iñigo MA. Dificultades quirúrgicas en el paciente

con enfermedad de paget: Revisión bibliográfica. Revista Española de Enfermedades Metabólicas Óseas. 2009;**18**(1):20-23. Available from: https://www.elsevier.es/es-revista-reemo-70-articulo-difucultades-quirurgicas-el-paciente-con-13138551

[38] Jara PL. Monografía: La fractura de cadera en el paciente mayor. La importancia del papel de la enfermera en la rehabilitación precoz. Trabajo Fin de Grado. Universidad de Cantabria; 2014. Available from: https://repositorio.unican.es/xmlui/bitstream/handle/10902/4625/%5B1%5D%20JaraPerezL.pdf?sequence=1

[39] San Pío TV, Lampré MPM, Gimeno ABE, Romero ACS, Clarín EF. Alteraciones de la marcha, inestabilidad y caidas. Tratado de geriatría para residentes. Madrid: Sociedad Española de Geriatría y Gerontología; 2007. Available from: http://www.segg.es/tratadogeriatria/PDF/S35-05%2019_II.pdf

[40] Pech-Ciau BA, Lima-Martínez EA, Espinosa-Cruz GA, Pacho-Aguilar CR, Huchim-Lara O, Alejos-Gómez RA. Fractura de cadera en el adulto mayor: epidemiología y costos de la atención. Acta Ortopédica Mexicana. 2021;**35**(4):341-347. DOI: 10.35366/103314

[41] Muñoz S, Lavanderos J, Vilches L. Fractura de cadera. Cuadernos de Cirugía (Valdivia). 2008;**22**(1):73-81. Available from: http://mingaonline.uach.cl/pdf/cuadcir/v22n1/art11.pdf

[42] Zaragoza JA, Portal LF. Guía de buena práctica clínica en geriatría. Anciano afecto de fractura de cadera. Sociedad Española de Geriatría y Gerontología. Barcelona: Sociedad Española de Cirugía Ortopédica y Elsevier Doyma; 2007. Available from: http://www.segg.es/sites/default/files/page/guia_fractura_cadera.pdf

[43] Zujur D, Álvarez-Barreto JF. Prótesis En Artroplastia Total De Cadera Y Recubrimientos Bioactivos De Quitosano Para Mejorar Su Desempeño. Revista Ingeniería Biomédica. 2016;**10**(19):33-43. Available from: http://www.scielo.org.co/scielo.php?script=sci_arttext&pid=S1909-97622016000100004&lng=en

Chapter 5

Fast-Track Rehabilitation Focusing on Nutritional Support during the Perioperative Period of Total Hip Arthroplasty

Takashi Ikeda, Kazunari Ninomiya, Satoshi Koto and Keigo Nanjo

Abstract

Total hip arthroplasty (THA) is a highly effective technique for relieving pain and reconstructing joint structures. However, even several years after THA, the preexisting muscle weakness does not resolve to the level of healthy individuals. Since the 2000s, minimally invasive surgical procedures and fast-track rehabilitation have enabled early functional recovery, particularly in terms of walking ability, but solutions to these problems have not yet been established. The benefits of combined nutrition and exercise interventions for sarcopenia and frailty are gaining widespread acceptance. Elements of sarcopenia and frailty may be inherently present in patients who have undergone THA, and a combination of nutritional and exercise interventions may be useful in treating post-prosthetic muscle weakness and prolonged muscle atrophy. This chapter describes their usefulness and implementation for patients who have undergone THA.

Keywords: nutritional support, muscle atrophy, nutritional physiology, total hip arthroplasty, muscle weakness

1. Introduction

Total hip arthroplasty (THA) is an excellent surgical technique to alleviate hip pain and improve activities of daily living (ADL) and quality of life (QOL) in patients with end-stage hip osteoarthritis, and the number of THAs performed worldwide has increased year after year [1]. Particularly, THA has helped achieve favorable long-term results due to innovations in prosthetic implants and surgical techniques in recent years [1, 2].

However, most patients undergoing THA develop muscle atrophy and weakness due to joint deformity and decreased physical activity [3, 4], require a prolonged postoperative functional recovery period [5, 6], and are at increased risk of serious complications, including falls [7], dislocations [8, 9], and infections [10]. In addition, prolonged muscle atrophy and weakness after surgery have negative consequences, such as a higher risk of frailty and care needs [11, 12], postoperative complications, and an

 IntechOpen

increased strain on social resources [13]. Therefore, there is an urgent need to understand the factors associated with muscle atrophy and weakness in patients who have undergone THA and to implement more effective interventions.

2. Functional and physiological changes in muscles before arthroplasty

Osteoarthritis (OA) and rheumatoid arthritis, for which THA is most commonly performed, are chronic diseases associated with changes in muscle function and quality due to both the disease and limb disuse. End-stage hip OA is associated with increased atrophy and fatty infiltration of the gluteus minor [13, 14], gluteus medius [13], and gluteus maximus muscles [14] in comparison to healthy individuals. The muscle mass of the gluteus maximus, gluteus medius, and gluteus minimus muscles in patients with end-stage OA showed 11.5, 6.9, and 13.7% asymmetry, respectively, compared to corresponding muscles in healthy individuals [14]. In the advanced stage of OA, a 12% asymmetry of the gluteus medius muscles has been reported as compared to healthy controls [15]. In addition, muscle biopsies of the quadriceps muscle have demonstrated atrophy of both type 2A and 2B muscle fibers [16]. A recent systematic review noted that muscle mass loss in patients with hip OA varied widely from muscle to muscle, while fatty infiltration was identified in several muscles [17, 18]. Against the background of these muscle mass and qualitative changes, the strength of the abductor muscle group is reduced compared to healthy subjects [13, 14, 17]. In terms of muscle activity, the gluteus minimus muscle in patients with OA is more active during the stance phase of walking compared to healthy individuals, with the degree of overactivity increasing with severity, while the gluteus medius muscle exhibits no difference [19]. On the other hand, it has been reported that there is an overall delay in muscle activity during stepping tasks, as well as increased activity in the gluteus minimus muscle, similar to walking [20].

Selective muscle atrophy of the vastus medialis muscle is observed in the early stages of knee OA [21]. Reduced quadriceps muscle strength is a risk factor for the progression of OA [22] and is 16–49% lower than in healthy older adults [21, 23, 24]. Clinical symptoms such as pain [25] and arthrogenic muscle inhibition (AMI) associated with joint edema [26] affect quadriceps muscle strength. Quadriceps muscle in knee OA patients exhibits more pronounced atrophy of type II fibers than type I fibers [27] and an increased proportion of non-contractile tissue within the muscle, such as fatty and connective tissue [21, 23, 24]. Maly et al. [28] found that females with advanced knee OA exhibit a greater percentage of fatty infiltration in the quadriceps muscle. Wada et al. [29] also reported that a one-kilogram increase in lower limb muscle mass for the joint protective effect could be expected to increase by 7.3 Nm in patients with early-stage knee OA, whereas it only increased by 3.8 Nm in end-stage knee OA. These reports confirm that changes in quadriceps muscle quality with the progression of knee OA affect muscle strength independently of muscle atrophy.

3. Dietary guidance and characteristics of dietary intake in patients with osteoarthritis

Dietary guidance for weight loss is widely used as a part of lifestyle guidance for osteoarthritis patients. However, obesity and metabolic syndrome are not risk factors for the progression of osteoarthritis [30, 31]. In contrast, Messier et al. [32, 33] reported that weight loss by restricting energy intake reduced joint pain and mechanical joint loading in knee OA patients; however, difficulties in maintaining energy intake restrictions have been noted for compliance and the long-term sustainability of weight loss. In addition, a recent scoping review suggested that the assessment for sarcopenic obesity should be included in osteoarthritis patients, as it is a risk factor for muscle weakness, loss of muscle mass, and post-prosthetic complications [34], and the risk of developing knee OA has been reported to increase approximately twofold [35]. With respect to necessary levels of protein, vitamins C, E, and omega-3 fatty acids, insufficient protein intake was observed to be more prevalent in older individuals with hip and knee osteoarthritis [36]. de Zwart et al. [37] also noted insufficient protein intake in most patients with knee OA. Additionally, there are also reports that a high intake of red meat is effective in reducing the need for THA [38]. Furthermore, adherence rates for the Mediterranean diet and dietary quality were significantly worse in OA patients [39], which suggests that a diet focused on anti-inflammatory effects may be beneficial [40].

4. Decrease in muscle mass at the perioperative period

The decrease in muscle mass after surgical procedures is partly due to accelerated catabolism caused by the surgical invasion and partly attributable to the muscle atrophy caused by rest. Surgical techniques for joint arthroplasty have developed rapidly, with mini-incision surgery techniques becoming more common since the 2000s, and surgical invasion and associated blood loss could be controlled [41]. Prevention of postoperative anemia is important because preoperative serum hemoglobin (Hb) is affected during the hospital stay after THA [42], and Hb decline after surgery affects lower limb muscle strength [43]. However, it was shown that moderate anemia had a limited impact on functional recovery [44].

In contrast, the effect of bed rest has been reported in patients with ankle fractures, where 7 days of bed rest with unloading resulted in a 6–16% reduction in the cross-sectional area of the quadriceps femoris muscle, whereas inpatients admitted to an intensive treatment unit experienced a 1.0–8.7% reduction over 3 days and an 8.8–13.7% reduction over 5 days [45]. In patients with hip fractures, type I muscle fibers did not differ among younger patients, while type II muscle fibers exhibited significant muscle atrophy compared to healthy older adults [46].

Enhanced Recovery After Surgery (ERAS) programs and early mobilization have become widespread in recent years and have been applied early in the area of arthroplasty. However, a systematic review [47] reported that employing ERAS after THA has no impact on functional improvement or complication prevention. This point should be considering Dreyer et al.'s [48] report of a decrease in muscle mass on both limbs on the operative and non-operative side after Total Knee Arthroplasty (TKA). Because patients requiring THA exhibit gluteal muscle atrophy before surgery, an assessment of postoperative muscle atrophy must also take into account the effects of increased bedrest, decreased loading of the lower limb that cannot be compensated by ERAS, selective atrophy of type II muscle fibers, and anemia due to blood loss.

5. Factors associated with medium- and long-term recovery in muscle strength and mass after THA

What will be the degree of functional improvement after early recovery and discharge by the ERAS program or fast-track surgery? Fukumoto et al. [49] found it to be lower than that of healthy controls [49, 50], although it did not reach preoperative levels at 1 month, but improved compared to preoperative values at 4 to 6 months, and hip abductor muscle strength was still affected by preoperative values at 6 months postoperatively. One year after THA, Judd et al. [51] reported that hip flexion and knee extension muscle strength were significantly lower in OA patients than in healthy individuals of the comparable age, whereas hip extension, abduction, and adduction muscle strength improved to the same level as healthy individuals of the same age. Recovery in muscle strength and mass up to 1 year after THA shows differences for each muscle group, reflecting differences in the site of muscle atrophy and fatty infiltration before THA (para 1). Furthermore, regarding the long-term course, it has also been reported that patients up to 6 years after THA did not achieve the level of healthy individuals of the comparable age [52], and similarly, hip abductor muscle strength remained significantly lower than in healthy individuals [53] after 10 years after THA.

Previous studies have shown that the hip and knee periprosthetic muscle strength after THA initially decreases compared to the level of preoperative strength up to about 1 month postoperatively but exceeds preoperative values at about 6 months postoperatively. Improvement thereafter is slower, and values are still lower than those of healthy individuals of comparable age more than 2 years after surgery. In other words, it can be inferred that in patients who have undergone THA, the hip and knee periprosthetic muscle weakness persists during the mid-to-long-term postoperative period. However, no consensus has been reached due to differences in surgical technique, femoral offset, and postoperative rehabilitation programs [54]. Furthermore, further research is needed on the medium- to long-term recovery of

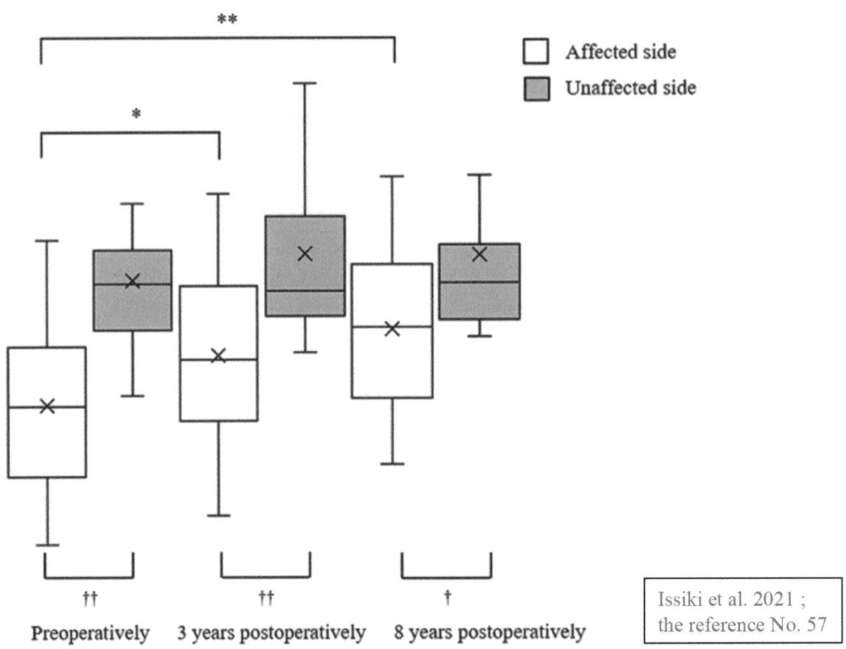

Figure 1.
Asymmetry of the cross-sectional area of the gluteus medius muscle. In the cross-sectional area of the gluteus medius muscle from preoperative to 8 year after THA, the muscle mass on the operated side was still significantly lower than on the healthy side at 8 years after THA, although it recovers over time. Source: Isshiki et al. [57].

periprosthetic hip muscle strength and related factors, as there are few reports on patients who are more than 2 years postoperative.

Several observational studies on changes in skeletal muscle mass after THA have been published, supporting reports of prolonged muscle weakness over the medium and long term. Reports up to 2 years after surgery show that the muscle mass of the affected side does not improve to the level of the healthy side in the whole gluteus muscle group [55, 56], while it has been reported that there is no difference in the ratio of its muscle mass between the healthy and affected side 1 year after surgery when limited to the gluteus medius muscle. Isshiki et al. [57] found that in patients with posterior lateral THA over 8 years postoperatively, the muscle cross-sectional area of the affected side did not improve to the level of the healthy side throughout the entire preoperative, 3-year, and 8-year postoperative periods (**Figure 1**). These findings suggest that atrophy of the periprosthetic hip muscle, mainly the gluteus medius muscle, in patients who have undergone THA patients may persist in the mid-to-long-term postoperatively. However, this is debatable due to differences in age, surgical technique, and postoperative rehabilitation [58–60]. Further studies are needed to adjust for influencing factors such as surgical technique and postoperative rehabilitation.

6. Sarcopenia-frail as a comorbidity after THA

Sarcopenia as a comorbidity has received increased attention in recent years. Regarding the prevalence of sarcopenia, Chang et al. [61] reported that of the 307,678 patients who underwent THA, 1319 patients (0.43%) had a prior diagnosis of sarcopenia, while Koto et al. [62] found it to be 8%, with differences in both reports, these prevalence ratios were less than the 47.3% of proximal femur fractures [63]. However, preoperative physical function was impaired in patients with concomitant sarcopenia [64], and they are at an increased risk of complications due to implant-related dislocation in the first year postoperatively [61]. Additionally, a higher incidence of falls and fragility fractures is observed [61, 65], resulting in higher readmission rates and healthcare costs [61]. At 1-year follow-up, patients with concomitant sarcopenia who had undergone THA were 62% more likely to fall and 77% more likely to develop fragility fractures [41], which suggests that falls and fragility fractures not only reduce the QOL but also increase patient mortality [65].

Frailty after THA has been reported to increase the incidence of postoperative complications, revision surgery, readmission, and mortality, as well as prolong hospital stays, as with sarcopenia, and ultimately increase healthcare costs [66–70]. Therefore, early diagnosis and prevention of frailty in patients who have received THA are urgent issues. We investigated the prevalence of frailty and associated factors in community-dwelling elderly patients 1 year following THA and found a prevalence of 11.2% for frailty and 51.0% for pre-frailty [11]. The results showed a higher percentage compared to the prevalence in Japanese community-dwelling older adults (frail: 8.7%, pre-frail: 40.8%). In addition, the history of falls, maximum leg circumference, hip abductor strength, knee extensor strength, and performance on the Timed up & go test were significantly associated with frailty in patients who had undergone THA. These findings suggest that improving muscle mass and dynamic balance capacity as well as increasing muscle strength, particularly hip abductor and knee extensor strength, may be important to prevent frailty after THA. However, there are few reports on the coexistence of osteoarthritis and frailty [71, 72], and the causal relationship with frailty is also unclear [73].

7. Consensus on therapeutic interventions for sarcopenia and frailty

Research on sarcopenia has progressed dramatically in the last decade, and in recent years' guidelines [74], consensus papers [75, 76], and position papers [77] have been successively published. The common view shared by these publications is that treatments can improve muscle strength and mass when combined with exercise and nutritional therapy. Muscle mass loss can occur under any of the conditions of sarcopenia, malnutrition, and cachexia [76], but each of these factors needs to be taken into account, as they are exacerbated by one another [77]. The recommended exercise intensity is 50% of 1 repetition maximum (1RM) and 80% of 1RM for muscle hypertrophy if feasible [78]. For nutritional therapy, a protein intake of 1–1.5 g per kg body weight per day is recommended, and there is insufficient evidence regarding the effectiveness of vitamin D [77]. The International Clinical Practice Guidelines for Sarcopenia [74] also note the importance of exercise therapy and protein-rich diet or supplementation, as well as education on these treatments.

Sarcopenia is also encompassed within the concept of frailty and overlaps with physical frailty in many aspects [79]. As with sarcopenia, several guidelines [80, 81] and recommendations [82] have been reported. As with sarcopenia, a combination of exercise and nutritional therapy is recommended, with exercise therapy encompassing resistance training and nutritional therapy consisting of protein and caloric supplementation [80–82]. In addition, vitamin D is recommended only in Asian regions [75], and cognitive or problem-solving therapy is not recommended [81]. Also, a recent recommendation has proposed screening for frailty in inpatients over 70 years of age [82].

8. Fast-track surgery and rehabilitation in THA

Fast-track surgery is a multimodal effort involving a multidisciplinary team of anesthetists, physiotherapists, and nurses in addition to the surgeon. Enhanced recovery requires a combination of preoperative education, stress reduction, pain relief, early ambulation and mobilization, drain and tube management and their removal to enable mobility, as well as fluid and nutritional therapy [83]. Fast-track surgery in THA, together with minimally invasive techniques, facilitates early functional recovery and discharge through multimodal pain management [84] and clinical nursing pathways [85]. This fast-track initiative was also reported to have no problems in terms of medical safety [86]. Compared to conventional perioperative management, fast-track surgery has been reported to reduce hospital stay [87], which is becoming more prevalent in patients undergoing THA, although there is less emphasis on nutritional support.

The main outcomes of fast-track surgery in THA are early mobility and discharge from the hospital. As described in the previous paragraphs, patients who have undergone THA exhibit the following characteristics: the preoperative presence of disuse changes in muscle function and qualitative aspects of physiology (par. 1), poor protein intake and adherence rate (par. 2), increased bedrest and decreased loading of the lower limb that cannot be compensated for even with the ERAS program, and selective atrophy of type II muscle fibers (par. 3). Additionally, prolonged muscle weakness in the mid-to-long-term postoperative period (par. 4) and the coexistence of frailty and sarcopenia in patients who received THA (par. 5) have also been identified, which have not been adequately addressed by initiatives such as fast-track

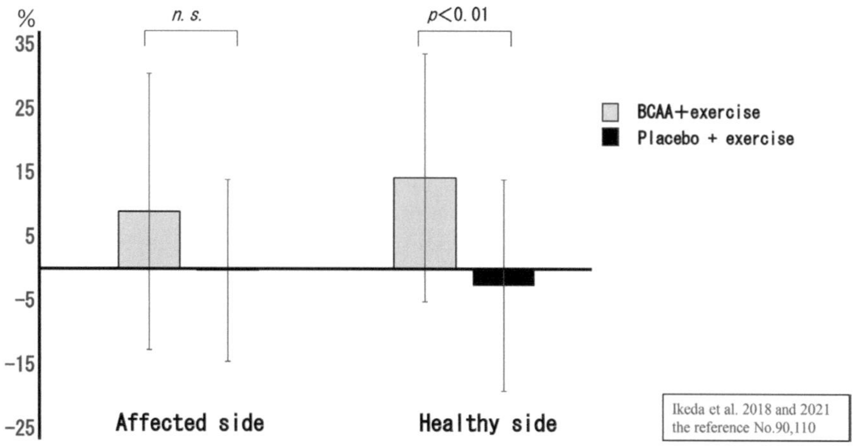

Figure 2.
Improvement rate in hip abductor strength (patients with hip OA). The BCAA group showed a greater rate of improvement on both sides and was significantly higher on the healthy side. Source: Ikeda et al. [89–110].

surgery and rehabilitation. As the consensus for the combination of exercise and nutritional therapy as a therapeutic intervention in frailty and sarcopenia grows, initiatives to incorporate nutritional therapy into the fast-track component of THA are emerging.

9. Combination of exercise and nutritional interventions and their effects before surgery

Resistance training in combination with high protein intake for patients with OA has been suggested in systematic reviews to be potentially useful in improving lower limb muscle strength and motor function [88]. In one of the reviews, Ikeda et al. [89] randomly provided 6 g of branched-chain amino acid (BCAA) for 4 weeks to patients with end-stage hip OA scheduled for THA. An improvement in hip abductor muscle strength on the healthy side was observed (**Figure 2**), but they reported no difference in the affected side's muscle strength compared to the placebo. One reason for this is that resistance exercise in combination with nutritional therapy has been reported to have no significant effect on improving muscle mass, muscle strength, and physical functions in healthy older people [90, 91]. In other words, the additional effect of nutritional supplementation is unlikely to be observed or limited in older people who were able to obtain sufficient amounts of nutrition, including protein, from their daily diet. As common challenges in OA patients, inadequate protein intake [36], and poor adherence to a Mediterranean diet are prevalent [39]. Future large-scale studies are needed to assess confounding factors such as age, nutritional status, daily diet, and physical activity of the participants and to adjust for these factors.

10. Combination of exercise and nutritional interventions and their effects after THA

In recent years, patients have been able to attain early mobility following THA, from the day of surgery due to remarkable advances in prosthetic implants and surgical techniques, which have also enabled the reduction of healthcare costs and hospital stays [92–94]. However, it has also been noted that approximately 70% of elderly inpatients for THA were discharged with the same degree of physical function as before surgery [95] and that prolonged functional decline is associated with muscle atrophy and weakness. Most of the patients who underwent THA exhibit inadequate nutritional intake preoperatively [36, 37]. This malnutrition not only prolongs functional recovery but also increases the incidence of postoperative complications, readmission rates, and hospital stays [61, 96]. In short, we believe that exercise therapy in combination with nutritional interventions, rather than exercise therapy alone, can maximize the induction of muscle protein synthesis, reduce muscle atrophy, and lead to a more efficient and early functional recovery.

Two randomized controlled trials (RCTs) in patients who underwent TKA and two in those who received THA have examined the influence of dietary intake on functional recovery following arthroplasty. In perioperative TKA patients, Ueyama et al. [97] conducted an RCT in which 3 g of essential amino acids (EAA) was provided three times daily from 1 week before to 2 weeks after surgery and investigated quadriceps muscle atrophy, knee extension muscle strength, and ADL ability up to 4 weeks after TKA. The authors reported that the intervention group showed significantly reduced postoperative muscle atrophy and early improvement in ADL functions. In addition, Dreyer et al. [48] conducted an RCT in perioperative patients planned for TKA and who were administered 20 g of EAA twice daily from 1 week before to 6 weeks after surgery to investigate quadriceps muscle atrophy and knee extensor strength up to 6 weeks after TKA. They reported significantly reduced postoperative muscle atrophy in the intervention group but no significant difference in knee extensor strength.

Ninomiya and Ikeda [98] compared 29 perioperative THA patients who were administered a high protein dietary supplement containing 3.4 g BCAA twice daily from 4 weeks before surgery to 8 weeks after surgery (BCAA group) with 29 patients who were administered a regular program (control group). They compared hip abductor strength and knee extensor strength between both groups for up to 8 weeks after surgery. The strength of hip abductors on the affected side did not differ between the two groups, whereas the strength of hip abductors on the healthy side and bilateral knee extensors was considerably greater in the BCAA group. In a study of patients who underwent THA and were in the convalescent rehabilitation period, Ikeda et al. [99] compared 18 patients (the BCAA group) who were provided with 3.4 g BCAA once daily for 1 month in conjunction with physiotherapy twice daily and 13 patients (the control group) who were provided with 1.2 g placebo with physiotherapy twice daily. They found that the BCAA prevented a decrease in skeletal muscle mass and reported significantly higher values for the knee extensor muscles on the operated side (**Figures 3** and **4**). However, no significant difference was observed in bilateral hip abduction muscle strength between both groups. These reports suggest that adequate nutritional intake in patients requiring THA or TKA may prevent postoperative muscle atrophy and have a limited effect on muscle strength and functional recovery, although this is debatable. This is because (1) AMI due to swelling and pain is associated with early postoperative muscle strength [26, 100–102]; (2) the effects of nutritional intake may be synergistic when combined with resistance training [103, 104], and additionally, the type, amount, and timing of nutrient intake may play a role. Future studies are needed to examine the effect of nutritional intake on functional recovery in patients undergoing THA after adjusting for these factors.

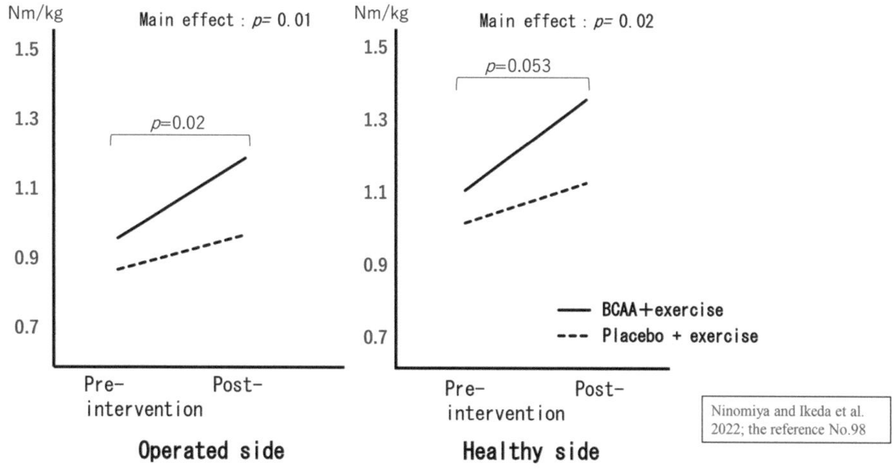

Figure 3.
Knee extensor strength pre-intervention and post-in-perioperative THA patients. There was no interaction between the two groups in terms of change in knee extensor strength. However, there was a significant main effect between groups and at time on the both side. Source: Ninomiya et al. [98–110].

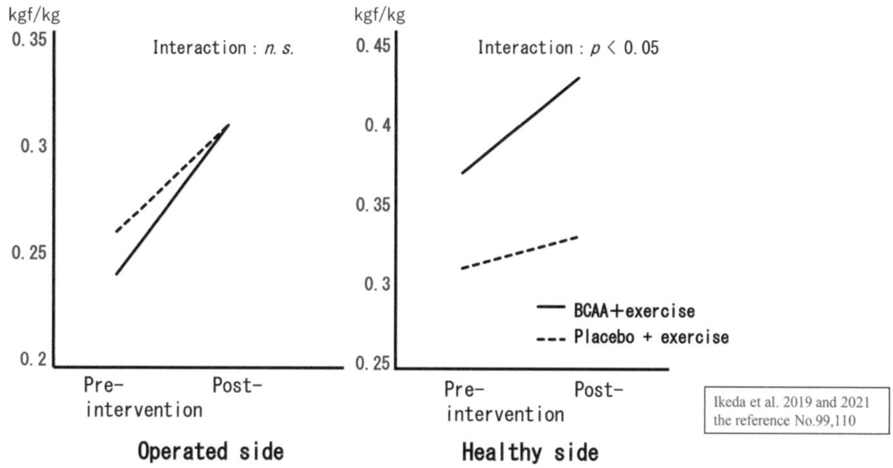

Figure 4.
Interaction plot of knee extensor strength pre-intervention and post-in convalescent THA patients. A two-way analysis of variance showed an interaction (synergy) between exercise and nutritional therapy in knee extensor strength on the healthy side. Source: Ikeda et al. [99–110].

11. Problems in preventing prolonged muscle weakness after THA

We have previously reported that limb muscle strength and motor function in patients 10 years after THA were significantly lower, and the incidence of falls was 2.8 times higher than in healthy individuals of comparable age [53]. The incidence of fractures resulting from falls was reported to be about twice as high as the incidence of implant-related postoperative complications (dislocation and wear) [105]. While THA in recent years has undergone dramatic advancements in prosthetic implants that have reduced the incidence of revision surgery due to dislocation and wear and increased implant survival rates, patients are likely to be at increased risk of requiring

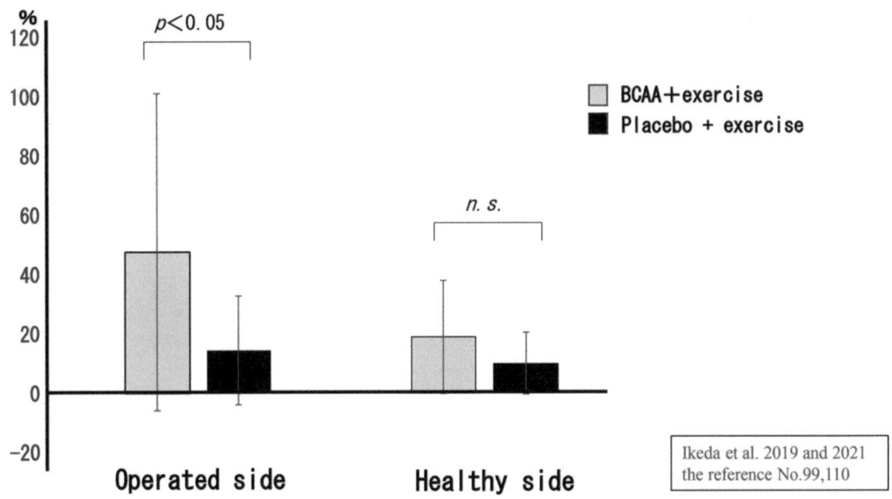

Figure 5.
Improvement rate in knee extensor strength (convalescent THA patients). The BCAA group showed that improvement rate in knee extensor strength was significantly higher on the healthy side. Source: Ikeda et al. [99–110].

care due to functional decline associated with age-related loss of muscle mass and strength. Currently, physiotherapy for patients undergoing THA includes exercise therapy and patient education with a focus on resistance training, with outcomes of improved lower limb muscle strength, walking ability, ADL functions, and QOL [106, 107]. However, recent studies have reported that physiotherapy after THA is as effective as independent training [108], pointing to the need to reconsider intervention methods. We reported that protein intake was associated with muscle strength in patients 1 year after THA [109]. Ueyama et al. [110] also reported that nutritional interventions were associated with enhanced muscle mass and strength in patients 2 years after THA. Therefore, to maintain and improve muscle mass, strength, and motor function in patients in the mid-to long-term following THA, it may be necessary not only to use exercise therapy but also to monitor the nutritional status of the patient and to use nutritional supplementation in combination (**Figure 5**).

12. Conclusion

Postoperative outcomes in patients undergoing THA have been highly successful in terms of early postoperative functional recovery and implant survival. We had assumed that the prolonged muscle weakness following THA would resolve with time. However, recent findings have indicated a situation that worsens with time. Fast-track rehabilitation, which focuses on a combination of exercise and nutritional therapy as a strategy to address preexisting muscle atrophy, fatty infiltration, and postoperative challenges, is a new technique that may provide a new tool for mitigating the problem of prolonged postoperative muscle weakness, which has remained unresolved for 20 years.

Acknowledgements

This book chapter writing was funded by Showa University Research Administration Center (Grant Number: 13000157).

Conflict of interest

The authors declare no conflict of interest.

Author details

Takashi Ikeda[1*], Kazunari Ninomiya[2], Satoshi Koto[1] and Keigo Nanjo[3]

1 School of Nursing and Rehabilitation Sciences, Showa University, Kanagawa, Japan

2 Shonan Kamakura Joint Reconstruction Center, Kanagawa, Japan

3 Center for Rehabilitation, Shonan Kamakura General Hospital, Kanagawa, Japan

*Address all correspondence to: tk.ikeda@nr.showa-u.ac.jp

References

[1] Fontalis A, Epinette JA, Thaler M, et al. Advances and innovations in total hip arthroplasty. SICOT Journal. 2021;7:26. DOI: 10.1051/sicotj/2021025

[2] Hopper RH Jr, Ho H, Sritulanondha S, et al. Crosslinking reduces THA Wear, osteolysis, and revision rates at 15-year Followup compared with noncrosslinked polyethylene. Clinical Orthopaedics and Related Research. 2018;**476**(2):279-290. DOI: 10.1007/s11999.0000000000000036

[3] Rasch A, Byström AH, Dalen N, et al. Reduced muscle radiological density, cross-sectional area, and strength of major hip and knee muscles in 22 patients with hip osteoarthritis. Acta Orthopaedica. 2007;**78**(4):505-510. DOI: 10.1080/17453670710014158

[4] Loureiro A, Constantinou M, Diamond LE, et al. Individuals with mild-to-moderate hip osteoarthritis have lower limb muscle strength and volume deficits. BMC Musculoskeletal Disorders. 2018;**19**(1):303. DOI: 10.1186/s12891-018-2230-4

[5] Nanri Y, Shibuya M, Nozaki K, et al. The impact of sarcopenia risk on postoperative walking independence in older adults undergoing total joint arthroplasty. Journal of Geriatric Physical Therapy. 2024;**47**(1):28-35. DOI: 10.1519/JPT.0000000000000368

[6] Nankaku M, Tsuboyama T, Aoyama T, et al. Preoperative gluteus medius muscle atrophy as a predictor of walking ability after total hip arthroplasty. Physical Therapy Research. 2016;**19**(1):8-12. DOI: 10.1298/ptr.e9884

[7] Ninomiya K, Takahira N, Ikeda T, et al. Predictors of falls in patients during the first year after total hip arthroplasty: A prospective cohort study. Health Science Reports. 2020;**3**(3):e184. DOI: 10.1002/hsr2.184

[8] García-Rey E, García-Cimbrelo E. Abductor biomechanics clinically impact the total hip arthroplasty dislocation rate: A prospective long-term study. The Journal of Arthroplasty. 2016;**31**(2):484-490. DOI: 10.1016/j.arth.2015.09.039

[9] Elbuluk AM, Coxe FR, Schimizzi GV, et al. Abductor deficiency-induced recurrent instability after Total hip arthroplasty. JBJS Reviews. 2020;**8**(1):e0164. DOI: 10.2106/JBJS.RVW.18.00164

[10] Babu JM, Kalagara S, Durand W, et al. Sarcopenia as a risk factor for prosthetic infection after Total hip or knee arthroplasty. The Journal of Arthroplasty. 2019;**34**(1):116-122. DOI: 10.1016/j.arth.2018.09.037

[11] Ninomiya K, Takahira N, Ikeda T, et al. Prevalence of frailty and associated factors among community-dwelling older adults after total hip arthroplasty. HIP International. 2023;**33**(3):397-403. DOI: 10.1177/11207000211048180

[12] Johnson RL, Frank RD, Abdel MP, et al. Frailty transitions one year after Total joint arthroplasty: A cohort study. The Journal of Arthroplasty. 2022;**37**(1):10-18.e2. DOI: 10.1016/j.arth.2021.08.022

[13] Kivle K, Lindland ES, Mjaaland KE, et al. Gluteal atrophy and fatty infiltration in end-stage osteoarthritis of the hip: A case-control study. Bone & Joint Open. 2021;**2-1**:40-47. DOI: 10.1302/2633-1462.21.BJO-2020-0179.R1

[14] Zacharias A, Pizzari T, English DJ, et al. Hip abductor muscle volume in hip osteoarthritis and matched controls. Osteoarthritis and Cartilage. 2016;**24**(10):1727-1735. DOI: 10.1016/j.joca.2016.05.002

[15] Grimaldi A, Richardson C, Stanton W, et al. The association between degenerative hip joint pathology and size of the gluteus medius, gluteus minimus and piriformis muscles. Manual Therapy. 2009;**14**:605-610. DOI: 10.1016/j.math.2009.07.004

[16] Reardon K, Galea M, Dennett X, et al. Quadriceps muscle wasting persists 5 months after total hip arthroplasty for osteoarthritis of the hip: A pilot study. Internal Medicine Journal. 2001;**31**(1):7-14. DOI: 10.1046/j.1445-5994.2001.00007.x

[17] Marshall AR, de Noronha M, Zacharias A, et al. Structure and function of the abductors in patients with hip osteoarthritis: Systematic review and meta-analysis. Journal of Back and Musculoskeletal Rehabilitation. 2016;**29**(2):191-204. DOI: 10.3233/BMR-150614

[18] Lawrenson PR, Crossley KM, Vicenzino BT, et al. Muscle size and composition in people with articular hip pathology: A systematic review with meta-analysis. Osteoarthritis and Cartilage. 2019;**27**:181-195. DOI: 10.1016/j.joca.2018.10.008

[19] Zacharias A, Pizzari T, Semciw AI, et al. Comparison of gluteus medius and minimus activity during gait in people with hip osteoarthritis and matched controls. Scandinavian Journal of Medicine & Science in Sports. 2019;**29**(5):696-705. DOI: 10.1111/sms.13379

[20] Zacharias A, Pizzari T, Semciw AI, et al. Gluteus medius and minimus activity during stepping tasks: Comparisons between people with hip osteoarthritis and matched control participants. Gait & Posture. 2020;**80**:339-346. DOI: 10.1016/j.gaitpost.2020.06.012

[21] Taniguchi M, Fukumoto Y, Kobayashi M, et al. Quantity and quality of the lower extremity muscles in women with knee osteoarthritis. Ultrasound in Medicine & Biology. 2015;**41**(10):2567-2574. DOI: 10.1016/j.ultrasmedbio.2015.05.014

[22] Øiestad BE, Juhl CB, Culvenor AG, et al. Knee extensor muscle weakness is a risk factor for the development of knee osteoarthritis: An updated systematic review and meta-analysis including 46 819 men and women. British Journal of Sports Medicine. 2022;**56**(6):349-355. DOI: 10.1136/bjsports-2021-104861

[23] Kumar D, Karampinos DC, MacLeod TD, et al. Quadriceps intramuscular fat fraction rather than muscle size is associated with knee osteoarthritis. Osteoarthritis and Cartilage. 2014;**22**(2):226-234. DOI: 10.1016/j.joca.2013.12.005

[24] Noehren B, Kosmac K, Walton RG, et al. Alterations in quadriceps muscle cellular and molecular properties in adults with moderate knee osteoarthritis. Osteoarthritis and Cartilage. 2018;**26**(10):1359-1368. DOI: 10.1016/j.joca.2018.05.011

[25] Riddle DL, Stratford PW. Impact of pain reported during isometric quadriceps muscle strength testing in people with knee pain: Data from the osteoarthritis initiative. Physical Therapy. 2011;**91**(10):1478-1489. DOI: 10.2522/ptj.20110034

[26] Rice DA, McNair PJ. Quadriceps arthrogenic muscle inhibition: Neural

mechanisms and treatment perspectives. Seminars in Arthritis and Rheumatism. 2010;**40**(3):250-266. DOI: 10.1016/j.semarthrit.2009.10.001

[27] Fink B, Egl M, Singer J, et al. Morphologic changes in the vastus medialis muscle in patients with osteoarthritis of the knee. Arthritis and Rheumatism. 2007;**56**(11):3626-3633. DOI: 10.1002/art.22960

[28] Maly MR, Calder KM, Macintyre NJ, et al. Relationship of intermuscular fat volume in the thigh with knee extensor strength and physical performance in women at risk of or with knee osteoarthritis. Arthritis Care & Research (Hoboken). 2013;**65**(1):44-52. DOI: 10.1002/acr.21868

[29] Wada O, Kurita N, Kamitani T, et al. Influence of the severity of knee osteoarthritis on the association between leg muscle mass and quadriceps strength: The SPSS-OK study. Clinical Rheumatology. 2019;**38**(3):719-725. DOI: 10.1007/s10067-018-4337-2

[30] Grotle M, Hagen KB, Natvig B, et al. Obesity and osteoarthritis in knee, hip and/or hand: An epidemiological study in the general population with 10 years follow-up. BMC Musculoskeletal Disorders. 2008;**9**:132. DOI: 10.1186/1471-2474-9-132

[31] Walter SS, Wintermeyer E, Klinger C, et al. Association between metabolic syndrome and hiposteoarthritis in middle-aged men and women from the general population. PLoS One. 2020;**15**(3):e0230185. DOI: 10.1371/journal.pone.0230185

[32] Messier SP, Beavers DP, Mihalko SL, et al. The effects of intensive dietary weight loss and exercise on gait in overweight and obese adults with knee osteoarthritis. The intensive diet and exercise for arthritis (IDEA) trial. Journal of Biomechanics. 2020;**98**:109477. DOI: 10.1016/j.jbiomech.2019.109477

[33] Messier SP, Mihalko SL, Legault C, et al. Effects of intensive diet and exercise on knee joint loads, inflammation, and clinical outcomes among overweight and obese adults with knee osteoarthritis: The IDEA randomized clinical trial. Journal of the American Medical Association. 2013;**310**(12):1263-1273. DOI: 10.1001/jama.2013.277669

[34] Godziuk K, Prado CM, Linda J, et al. The impact of sarcopenic obesity on knee and hip osteoarthritis: A scoping review. BMC Musculoskeletal Disorders. 2018;**19**:271. DOI: 10.1186/s12891-018-2175-7

[35] Hall KD, Kahan S. Maintenance of lost weight and long-term management of obesity. The Medical Clinics of North America. 2018;**102**(1):183-197. DOI: 10.1016/j.mcna.2017.08.012

[36] Purcell S, Thornberry R, Elliott SA, et al. Body composition, strength, and dietary intake of patients with hip or knee osteoarthritis. Canadian Journal of Dietetic Practice and Research. 2016;**77**(2):98-102. DOI: 10.3148/cjdpr-2015-037

[37] de Zwart AH, van der Leeden M, Roorda LD, et al. Dietary protein intake and upper leg muscle strength in subjects with knee osteoarthritis data from the osteoarthritis initiative. Rheumatology International. 2019;**39**(2):277-284. DOI: 10.1007/s00296-018-4223-x

[38] Wang Y, Simpson JA, Wluka EA, et al. Meat consumption and risk of primary hip and knee joint replacement due to osteoarthritis: A prospective cohort study. BMC Musculoskeletal Disorders. 2011;**12**:17. DOI: 10.1007/s00296-018-4223-x

[39] Comee L, Taylor CA, Nahikian-Nelms M, et al. Dietary patterns and nutrient intake of individuals with rheumatoid arthritis and osteoarthritis in the United States. Nutrition. 2019;**67-68**:110533. DOI: 10.1016/j.nut.2019.06.014

[40] Sadeghi A, Zarrinjooiee G, Mousavi SN, et al. Effects of a Mediterranean diet compared with the low-fat diet on patients with knee osteoarthritis: A randomized feeding trial. International Journal of Clinical Practice. 2022;**2022**:7275192. DOI: 10.1155/2022/7275192

[41] Xu CP, Li X, Song JQ, et al. Mini-incision versus standard incision Total hip arthroplasty regarding surgical outcomes: A systematic review and meta-analysis of randomized controlled trials. PLoS One. 2013;**8**(11):e80021. DOI: 10.1371/journal.pone.0080021

[42] Sicat CS, Muthusamy N, Singh V, et al. Impact of preoperative Anemia severity on primary Total hip arthroplasty outcomes. The Journal of Arthroplasty. 2022;**37**:721-726. DOI: 10.1016/j.arth.2021.12.042

[43] Maezawa K, Nozawa M, Yuasa T, et al. Postoperative hemoglobin and recovery of hip muscle strength after total hip arthroplasty. Journal of Orthopaedics. 2018;**15**:886-888. DOI: 10.1016/j.jor.2018.08.016

[44] Jans Ø, Bandholm T, Kurbegovic S, et al. Postoperative anemia and early functional outcomes after fast-track hip arthroplasty: A prospective cohort study. Transfusion. 2016;**56**(4):917-925. DOI: 10.1111/trf.13508

[45] Hardy EJO, Inns TB, Hatt J, et al. The time course of disuse muscle atrophy of the lower limb in health and disease. Journal of Cachexia, Sarcopenia and Muscle. 2022;**13**:2616-2629. DOI: 10.1002/jcsm.13067

[46] Kramer IF, Snijders T, Smeets JSJ, et al. Extensive type II muscle Fiber atrophy in elderly female hip fracture patients. The Journals of Gerontology. Series A, Biological Sciences and Medical Sciences. 2017;**72**(10):1369-1375. DOI: 10.1093/gerona/glw253

[47] Morrell AT, Daniel R, et al. Enhanced recovery after primary Total hip and knee arthroplasty: A systematic review. The Journal of Bone and Joint Surgery. American Volume. 2021;**103**:1938-1947. DOI: 10.2106/JBJS.20.02169

[48] Dreyer HC, Owen EC, Strycker LA, et al. Essential amino acid supplementation mitigates muscle atrophy after Total knee arthroplasty: A randomized, double-blind, placebo-controlled trial. The Journal of Bone and Joint Surgery. 2018:e0006. DOI: 10.2106/JBJS.OA.18.00006

[49] Fukumoto Y, Ohata K, Tsukagoshi R, et al. Changes in hip and knee muscle strength in patients following total hip arthroplasty. Journal of the Japanese Physical Therapy Association. 2013;**16**(1):22-27. DOI: 10.1298/jjpta.Vol16_002

[50] Klausmeier V, Lugade V, Jewett BA, et al. Is there faster recovery with an anterior or anterolateral THA? A pilot study. Clinical Orthopaedics and Related Research. 2010;**468**(2):533-541. DOI: 10.1007/s11999-009-1075-4

[51] Judd DL, Dennis DA, Thomas AC, et al. Muscle strength and functional recovery during the first year after THA. Clinical Orthopaedics and Related Research. 2014;**472**(2):654-664. DOI: 10.1007/s11999-013-3136-y

[52] Sicard-Rosenbaum L, Light KE, Behrman AL. Gait, lower extremity

strength, and self-assessed mobility after hip arthroplasty. The Journals of Gerontology. Series A, Biological Sciences and Medical Sciences. 2002;**57**(1):M47-M51. DOI: 10.1093/gerona/57.1.m47

[53] Ninomiya K, Hirakawa K, Ikeda T, et al. Patients 10 years after total hip arthroplasty have the deficits in functional performance, physical activity, and high fall rate compared to healthy adults. Physical Therapy Research. 2018;**21**(2):53-58. DOI: 10.1298/ptr.E9941

[54] Ismailidis P, Kvarda P, Vach W, et al. Abductor muscle strength deficit in patients after Total hip arthroplasty: A systematic review and meta-analysis. The Journal of Arthroplasty. 2021;**36**(8):3015-3027

[55] Rasch A, Byström AH, Dalén N, et al. Persisting muscle atrophy two years after replacement of the hip. Journal of Bone and Joint Surgery. British Volume (London). 2009;**91**(5):583-588

[56] Kawano T, Nankaku M, Murao M, et al. Recovery of muscle atrophy and fatty infiltration in patients with acetabular dysplasia after Total hip arthroplasty. Journal of the American Academy of Orthopaedic Surgeons. 2022;**30**(3):e317-e326

[57] Isshiki K, Jinno T, Aizawa J, et al. Asymmetry of the cross-sectional area of the gluteus Medius muscle persists eight years after Total hip arthroplasty for osteoarthritis of the hip. Progress in Rehabilitation Medicine. 2021;**25**(6):20210052. DOI: 10.2490/prm.20210052

[58] Müller M, Tohtz S, Dewey M, et al. Age-related appearance of muscle trauma in primary total hip arthroplasty and the benefit of a minimally invasive approach for patients older than 70 years. International Orthopaedics. 2011;**35**(2):165-171. DOI: 10.1007/s00264-010-1166-6

[59] Lalevée M, Curado J, Matsoukis J, et al. Comparative MRI assessment of three minimally invasive approaches in total hip arthroplasty. Orthopaedics & Traumatology, Surgery & Research. 2022;**108**(6):103354. DOI: 10.1016/j.otsr.2022.103354

[60] Rykov K, Meys TWGM, Knobben BAS, et al. MRI assessment of muscle damage after the posterolateral versus direct anterior approach for THA (Polada trial). A Randomized Controlled Trial, Journal of Arthroplasty. 2021;**36**(9):3248-3258.e1. DOI: 10.1016/j.arth.2021.05.00

[61] Chang KJ, Albright A, Testa EJ, et al. Sarcopenia is associated with an increased risk of postoperative complications following Total hip arthroplasty for osteoarthritis. Biology. 2023;**12**(2):295. DOI: 10.3390/biology12020295

[62] Koto S, Ikeda T, Inoue S, et al. Differences in preoperative function and outcome of patients with versus without sarcopenia after total hip arthroplasty. Journal of Physical Therapy Science. 2022;**34**(1):60-64. DOI: 10.1589/jpts.34.60

[63] Hida T, Ishiguro N, Shimokata H, et al. High prevalence of sarcopenia and reduced leg muscle mass in Japanese patients immediately after a hip fracture. Geriatrics & Gerontology International. 2013;**13**(1):413-420. DOI: 10.1111/j.1447-0594.2012.00918.x

[64] Ueoka K, Kabata T, Kajino Y, et al. The prevalence and impact of sarcopenia in females undergoing total hip arthroplasty: A

prospective study. Modern Rheumatology. 2022;**32**(1):193-198. DOI: 10.1080/14397595.2021.1899603

[65] Brown JP, Adachi JD, Schemitsch E, et al. Mortality in older adults following a fragility fracture: Real-world retrospective matched-cohort study in Ontario. BMC Musculoskeletal Disorders. 2021;**22**:105. DOI: 10.1186/s12891-021-03960-z

[66] Johnson RL, Abdel MP, Frank RD, et al. Impact of frailty on outcomes after primary and revision Total hip arthroplasty. The Journal of Arthroplasty. 2019;**34**(1):56-64.e5. DOI: 10.1016/j.arth.2018.09.078

[67] Wen H, Liu T, Li J. Association between frailty and clinical post-operative outcomes in patients following hip arthroplasty: A systematic review and meta-analysis. International Orthopaedics. 2023;**47**(3):667-675. DOI: 10.1007/s00264-022-05657-x

[68] Cook MJ, Lunt M, Ashcroft DM, et al. The impact of frailty on patient-reported outcomes following hip and knee arthroplasty. Age and Ageing. 2022;**51**(12):afac288. DOI: 10.1093/ageing/afac288

[69] Bai Y, Zhang XM, Sun X, et al. The association between frailty and mortality among lower limb arthroplasty patients: A systematic review and meta-analysis. BMC Geriatrics. 2022;**22**(1):702. DOI: 10.1186/s12877-022-03369-w

[70] Tram MK, O'Connor CM, Qian AS, et al. Frailty is associated with increased 30-day adverse events and hospitalization costs after primary Total hip arthroplasty. The Journal of Arthroplasty. 2022;**37**(8S):S925-S930. DOI: 10.1016/j.arth.2022.01.047

[71] Oosting E, Jans MP, Dronkers JJ, et al. Preoperative home-based physical therapy versus usual care to improve functional health of frail older adults scheduled for elective total hip arthroplasty: A pilot randomized controlled trial. Archives of Physical Medicine and Rehabilitation. 2012;**93**(4):610-616. DOI: 10.1016/j.apmr.2011.11.006

[72] Hoogeboom TJ, Dronkers JJ, van den Ende CH, et al. Preoperative therapeutic exercise in frail elderly scheduled for total hip replacement: A randomized pilot trial. Clinical Rehabilitation. 2010;**24**(10):901-910. DOI: 10.1177/0269215510371427

[73] O'Brien MS, McDougall JJ. Age and frailty as risk factors for the development of osteoarthritis. Mechanisms of Ageing and Development. 2019;**0**:21-28. DOI: 10.1016/j.mad.2019.03.003

[74] Dent E, Morley JE, Cruz-Jentoft AJ, et al. International clinical practice guidelines for sarcopenia (ICFSR): Screening, diagnosis and management. The Journal of Nutrition, Health & Aging. 2018;**22**(10):1148-1161. DOI: 10.1007/s12603-018-1139-9

[75] Chen LK, Woo J, Assantachai P, et al. Asian working Group for Sarcopenia: 2019 consensus update on sarcopenia diagnosis and treatment. Journal of the American Medical Directors Association. 2020;**21**(3):300-307.e2. DOI: 10.1016/j.jamda.2019.12.012

[76] Sayer AA, Cruz-Jentoft A. Sarcopenia definition, diagnosis and treatment: Consensus is growing. Age and Ageing. 2022;**51**(10):afac220. DOI: 10.1093/ageing/afac220

[77] Bauer J, Morley JE, Schols AMWJ, et al. Sarcopenia: A time for action. An SCWD Position Paper. Journal of Cachexia, Sarcopenia and Muscle. 2019;**10**(5):956-961. DOI: 10.1002/jcsm.12483

[78] Beckwée D, Delaere A, Aelbrecht S, et al. Exercise interventions for the prevention and treatment of sarcopenia. A systematic umbrella review. The Journal of Nutrition, Health & Aging. 2019;**23**(6):494-502. DOI: 10.1007/s12603-019-1196-8

[79] Cruz-Jentoft AJ, Sayer AA. Sarcopenia. Lancet. 2019;**393**(10191): 2636-2646. DOI: 10.1016/S0140-6736(19)31138-9

[80] Dent E, Lien C, Lim WS, et al. The Asia-Pacific clinical practice guidelines for the Management of Frailty. Journal of the American Medical Directors Association. 2017;**18**(7):564-575. DOI: 10.1016/j.jamda.2017.04.018

[81] Dent E, Morley JE, Cruz-Jentoft AJ, et al. Physical frailty: ICFSR international clinical practice guidelines for identification and management. The Journal of Nutrition, Health & Aging. 2019;**23**(9):771-787. DOI: 10.1007/s12603-019-1273-z

[82] Daly RM, Iuliano S, Fyfe JJ, et al. Screening, diagnosis and Management of Sarcopenia and Frailty in hospitalized older adults: Recommendations from the Australian and new Zealand Society for Sarcopenia and Frailty Research (ANZSSFR) expert working group. The Journal of Nutrition, Health & Aging. 2022;**26**(6):637-651. DOI: 10.1007/s12603-022-1801-0

[83] Kehlet H, Wilmore DW. Evidence-based surgical care and the evolution of fast-track surgery. Annals of Surgery. 2008;**248**(2):189-198. DOI: 10.1097/SLA.0b013e31817f2c1a

[84] Gaffney CJ, Pelt CE, Gililland JM, et al. Perioperative pain Management in hip and Knee Arthroplasty. The Orthopedic Clinics of North America. 2017;**48**(4):407-419. DOI: 10.1016/j.ocl.2017.05.001

[85] Zhang C, Xiao J. Application of fast-track surgery combined with a clinical nursing pathway in the rehabilitation of patients undergoing total hip arthroplasty. The Journal of International Medical Research. 2020;**48**(1):300060519889718

[86] Schneider M, Kawahara I, Ballantyne G, et al. Predictive factors influencing fast track rehabilitation following primary total hip and knee arthroplasty. Archives of Orthopaedic and Trauma Surgery. 2009;**129**(12):1585-1591. DOI: 10.1007/s00402-009-0825-9

[87] de Carvalho Almeida RF, Serra HO, de Oliveira LP. Fast-track versus conventional surgery in relation to time of hospital discharge following total hip arthroplasty: A single-center prospective study. Journal of Orthopaedic Surgery and Research. 2021;**16**(1):488. DOI: 10.1186/s13018-021-02640-x

[88] Liao CD, Wu YT, Tsauo JY, et al. Effects of protein supplementation combined with exercise training on muscle mass and function in older adults with lower-extremity osteoarthritis: A systematic review and meta-analysis of randomized trials. Nutrients. 2020;**12**(8):2422. DOI: 10.3390/nu12082422

[89] Ikeda T, Jinno T, Masuda T, et al. Effect of exercise therapy combined with branched-chain amino acid supplementation on muscle strengthening in persons with osteoarthritis. Hong Kong Physiotherapy Journal. 2018;**38**(1):23-31. DOI: 10.1142/S1013702518500038

[90] Labata-Lezaun N, Llurda-Almuzara L, López-de-Celis C, et al. Effectiveness of protein supplementation combined with resistance training on muscle strength and physical

performance in elderly: A systematic review and meta-analysis. Nutrients. 2020;**12**(9):2607. DOI: 10.3390/nu12092607

[91] Ten Haaf DSM, Nuijten MAH, Maessen MFH, et al. Effects of protein supplementation on lean body mass, muscle strength, and physical performance in nonfrail community-dwelling older adults: A systematic review and meta-analysis. The American Journal of Clinical Nutrition. 2018;**108**(5):1043-1059. DOI: 10.1093/ajcn/nqy192

[92] Skoffer B, Dalgas U, Mechlenburg I. Progressive resistance training before and after total hip and knee arthroplasty: A systematic review. Clinical Rehabilitation. 2015;**29**(1):14-29. DOI: 10.1177/0269215514537093

[93] Temporiti F, Draghici I, Fusi S, et al. Does walking the day of total hip arthroplasty speed up functional independence? A non-randomized controlled study. Archives of Physiotherapy. 2020;**24**(10):8. DOI: 10.1186/s40945-020-00079-7

[94] Büttner M, Mayer AM, Büchler B, et al. Economic analyses of fast-track total hip and knee arthroplasty: A systematic review. European Journal of Orthopaedic Surgery and Traumatology. 2020;**30**(1):67-74. DOI: 10.1007/s00590-019-02540-1

[95] Boyd CM, Landefeld CS, Counsell SR, et al. Recovery of activities of daily living in older adults after hospitalization for acute medical illness. Journal of the American Geriatrics Society. 2008;**56**(12):2171-2179. DOI: 10.1111/j.1532-5415.2008.02023.x

[96] Weijzen MEG, Kouw IWK, Verschuren AAJ, et al. Protein intake falls below 0.6 g•kg-1•d-1 in healthy, older patients admitted for elective hip or knee arthroplasty. The Journal of Nutrition, Health & Aging. 2019;**23**(3):299-305. DOI: 10.1007/s12603-019-1157-2

[97] Ueyama H, Kanemoto N, Minoda Y, et al. Perioperative essential amino acid supplementation suppresses rectus femoris muscle atrophy and accelerates early functional recovery following total knee arthroplasty. The Bone & Joint Journal. 2020;**102-B**:10-18. DOI: 10.1302/0301-620X.102B6.BJJ-2019-1370.R1

[98] Ninomiya K, Takahira N, Ikeda T, et al. Effects of perioperative exercise therapy combined with nutritional supplementation on functional recovery after fast-track total hip arthroplasty. Journal of Orthopaedic Science. DOI: 10.1016/j.jos.2022.09.012

[99] Ikeda T, Matsunaga Y, Kanbara M, et al. Effect of exercise therapy combined with branched-chain amino acid supplementation on muscle strength in elderly women after total hip arthroplasty: A randomized controlled trial. Asia Pacific Journal of Clinical Nutrition. 2019;**28**(4):720-726. DOI: 10.6133/apjcn.201912_28(4).0007

[100] Freeman S, Mascia A, McGill S. Arthrogenic neuromusculature inhibition: A foundational investigation of existence in the hip joint. Clinical Biomechanics (Bristol, Avon). 2013;**28**(2):171-177

[101] Lewis GN, Rice DA, McNair PJ, et al. Predictors of persistent pain after total knee arthroplasty: A systematic review and meta-analysis. British Journal of Anaesthesia. 2015;**114**(4):551-561. DOI: 10.1093/bja/aeu441

[102] Pua YH. The time course of knee swelling post Total knee arthroplasty and its associations with quadriceps

strength and gait speed. The Journal of Arthroplasty. 2015;**30**(7):1215-1219

[103] Kerksick CM, Arent S, Schoenfeld BJ, et al. International society of sports nutrition position stand: Nutrient timing. Journal of the International Society of Sports Nutrition. 2017;**14**:33. DOI: 10.1186/s12970-017-0189-4

[104] Stokes T, Hector AJ, Morton RW, et al. Recent perspectives regarding the role of dietary protein for the promotion of muscle hypertrophy with resistance exercise training. Nutrients. 2018;**10**(2):180. DOI: 10.3390/nu10020180

[105] Ninomiya K, Takahira N, Ochiai S, et al. Incidence of postoperative complications and non- periprosthetic fractures after total hip arthroplasty: A more than 10-year follow-up retrospective cohort study. Physical Therapy Research. 2020;**24**(1):77-83. DOI: 10.1298/ptr.E10043

[106] Moyer R, Ikert K, Long K, et al. The value of preoperative exercise and education for patients undergoing Total hip and knee arthroplasty: A systematic review and meta-analysis. JBJS Reviews. 2017;**5**(12):e2. DOI: 10.2106/JBJS.RVW.17.00015

[107] Saueressig T, Owen PJ, Zebisch J, et al. Evaluation of exercise interventions and outcomes after hip arthroplasty: A systematic review and meta-analysis. JAMA Network Open. 2021;**4**(2):e210254. DOI: 10.1001/jamanetworkopen.2021.0254

[108] Austin MS, Urbani BT, Fleischman AN, et al. Formal physical therapy after Total hip arthroplasty is not required: A randomized controlled trial. The Journal of Bone and Joint Surgery. American Volume. 2017;**99**(8):648-655. DOI: 10.2106/JBJS.16.00674

[109] Ikeda T, Sato R, Ninomiya K, et al. Nutritional factors related to muscle weakness 1 year after Total hip arthroplasty: A case-control study. Annals of Nutrition & Metabolism. 2022;**78**(2):73-79. DOI: 10.1159/000520551

[110] Jinno T, Ikeda T. Update of exercise therapy for osteoarthritis of the hip and total hip arthroplasty. Orthopaedic Surgery and Traumatology. 2021:427-433. DOI: 10.18888/se.0000001681

Chapter 6

Autologous Fat Transfer for Finger Joint and Basal Thumb Osteoarthritis

Marco Guidi, Lucian Lior Marcovici, Jaime Sánchez Ruas, Camilo Chaves, Fabio Andrés Tandioy-Delgado, Ahlam Arnaout, Carlos Federico Morales Hackett, Stefano Lucchina, Enrico Carità and Ivan Tami

Abstract

Autologous fat injection into osteoarthritic joints (liparthroplasty) has shown high potential in reducing pain and delaying the need for surgical intervention. The liparthroplasty is an alternative and minimally invasive treatment approach for finger and thumb carpometacarpal joint osteoarthritis with only a few studies available. A systematic literature review was performed with a search strategy in MEDLINE databases Google Scholar and Cochrane Library using the following keywords: "autologous fat grafting injection", "adipose-derived stem cells", "finger osteoarthritis", "basal thumb ostheoarthritis", "lipofilling". The database was analyzed from inception to August 1st, 2023. About 148 studies were identified; 17 additional articles were found through previous publications (total 165 articles). A total of 150 records resulted after duplicates were removed. Fourteen studies were selected and only 10 respected the inclusions criteria. In all these studies, intra-articular injection of autologous fat into osteoarthritic finger joints and trapeziometacarpal joint reduced pain and improved hand function. The liparthroplasty of osteoarthritic finger joints seems to be a minimally invasive, secure, and valid substitute for traditional surgical interventions. While the outcomes are promising, the lack of a standardized technique to fat processing and the absence of randomized controlled trials impede a thorough assessment of the procedure's effectiveness.

Keywords: autologous fat grafting, adipose-derived stem cells, basal thumb osteoarthritis, rhizoarthrosis, finger osteoarthritis, regenerative medicine, joint lipofilling, liparthroplasty

1. Introduction

Trapeziometacarpal (TM) and finger joint osteoarthritis are a progressive and disabling pathology of the hand [1]. Treatment can be conservative with orthosis, occupational therapy, anti-inflammatory drugs and intra-articular injections with cortisone and hyaluronic acid, resulting in only temporary success [1]. Furthermore, the time- and dose-dependent detrimental effects of the corticosteroids on articular cartilage *in vivo* and *in vitro* models are deeply documented in the literature [2–5]. For advanced osteoarthritis stages, different surgical techniques can be used such as denervation, arthrodesis, partial resection arthroplasty, trapeziectomy with or without suspensioplasty and hemi or total arthroplasty [6, 7]. However, there is considerable variability in surgeons' preferences, and none of the available techniques has demonstrated clear superiority over the others [8]. Studies on adipose tissue show its regenerative, anti-inflammatory and immunomodulatory properties [9–14]. Starting from the macrofat and nanofat grafting to the enzymatically separated vascular stromal fraction (SVF) and cultivated stem cells, adipose tissue has been the subject of extensive research in various degenerative diseases including knee and hip osteoarthritis [9]. However, the use of autologous fat grafting in treating basal thumb osteoarthritis and finger osteoarthritis is limited, with only a few studies available on this topic. Given the widespread interest in adipose tissue and its products with regenerative properties, this chapter was conducted to comprehensively assess their current clinical utility in treating osteoarthritis of finger joints. The aim of our investigation is to establish this technique as a reliable regenerative treatment option for finger and TM joint osteoarthritis.

2. Materials and methods

The systematic review was delineated following the PRISMA statement [15], using the guidelines described in the Cochrane Handbook for Systematic Reviews of Interventions.

2.1 Data sources and searches

A comprehensive search strategy was defined in collaboration with an independent research librarian and structured to capture all articles relating to autologous fat injection in osteoarthritic finger joints. The search strategy was applied to the MEDLINE databases, Google Scholar and Cochrane Library using the following keywords: "autologous fat grafting injection", "adipose-derived stem cells", "finger osteoarthritis", "basal thumb osteoarthritis", "lipofilling". The database was analyzed from inception to August 1st, 2023.

2.2 Inclusion/exclusion criteria

The inclusion and exclusion criteria were defined during the protocol stage. An article was included if the following criteria were fulfilled: the study was a primary study, case report or new technical report; the treatment was an injection of autologous fat with the description of harvesting technique and fat processing technique; follow-up, gender and age of the patients were reported; only studies with a design classification of evidence levels I-V were included.

Data not published on Pubmed, Google Scholar, Cochrane Library literature reviews, letters to editors, book chapters and presentations to congress were also excluded.

2.3 Selection of studies

The titles and abstracts of the studies were independently screened by two authors (MG, SL). The references of all studies and all review articles on the subject were also examined to identify other additional data to be included. This was subsequently followed by the assessment of the full text of the selection of studies. A third reviewer (IT) was helpful for any disagreements that could not be solved by this initial review.

2.4 Data extraction

Two reviewers (MG and SL) analyzed the data on an independent basis. The studies were scored for author name, year of publication, journal, population type and demographics and objective outcome measures.

2.5 Surgical technique

Many different techniques are described. The authors' preferred technique takes place under local anesthesia at the affected joint (using 5 ml of 1% lidocaine). A tumescent solution is employed for liposuction in the lower abdomen (upper thigh or lateral-gluteal region can also be used as donor site). The tumescent solution is prepared in sterile conditions, with a mixture of 100 ml of sodium chloride solution, 1 ml of 8.4% sodium bicarbonate and 10 ml of 1% lidocaine combined with 5 mcg/ml epinephrine. Roughly 50 ml of this solution is then evenly dispersed within the subcutaneous tissue.

Following a 10-min interval, a 2 mm incision is made in the lower portion of the umbilicus. By using a hemostat forceps, the subcutaneous tissue is dissected, enabling the insertion of a suction cannula (**Figure 1**). Approximately 15 ml of aspirate is

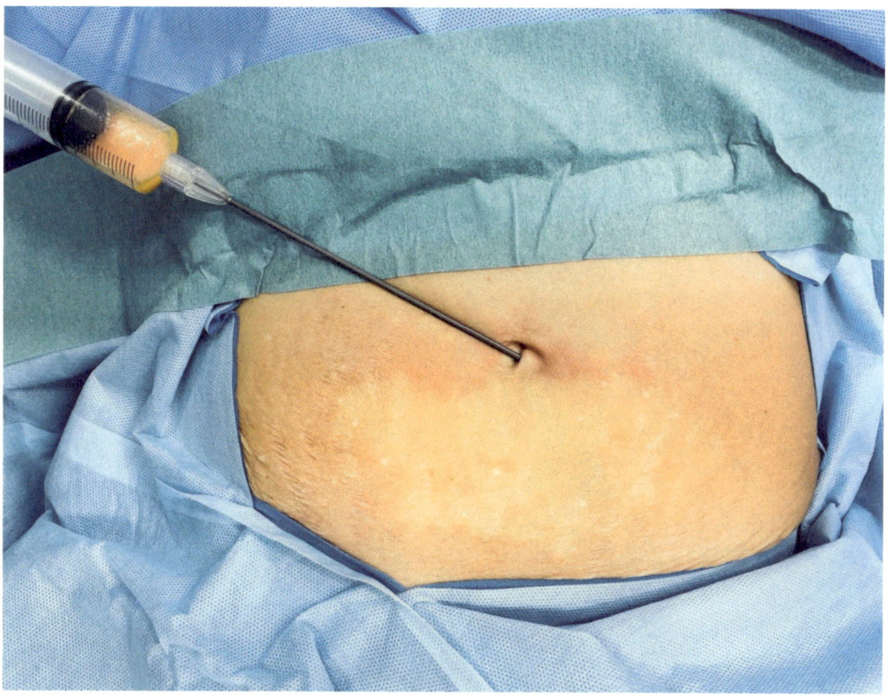

Figure 1.
A single umbilical incision was used. The lower abdomen skin was infiltrated with 50 mL solution (600 mg of lidocaine, 1000 mL of saline solution and 1 mL of 1:200,000 adrenaline). The vasoconstrictive effect of adrenalin is clearly visible by hypoperfusion of the skin.

collected and divided into two Luer-lock syringes (**Figure 2**). A variable amount of 0.5 to 1 ml of adipose tissue is introduced into each joint in a sterile field. To facilitate the insertion of the needle, a slight axial manual traction is applied on the finger, favoring the expansion of the joint space of the interphalangeal, metacarpophalangeal or trapeziometacarpal joint.

Depending on the size of each joint treated, fat tissue is injected in the amount that could be injected into the joint without over-pressure. Metacarpophalangeal joints receive a volume of adipose tissue of 1 ml, while the distal interphalangeal joints receive a smaller volume (0.5 ml) (**Figures 3** and **4**).

The adipose tissue is injected under fluoroscopy guidance. In multiple joint infiltrations, the needle must be changed for each infiltrated joint (**Figures 5** and **6**).

Subsequently, a thermoplastic splint is applied for the following 2 weeks. All patients undergo a functional recovery protocol with the occupational therapist.

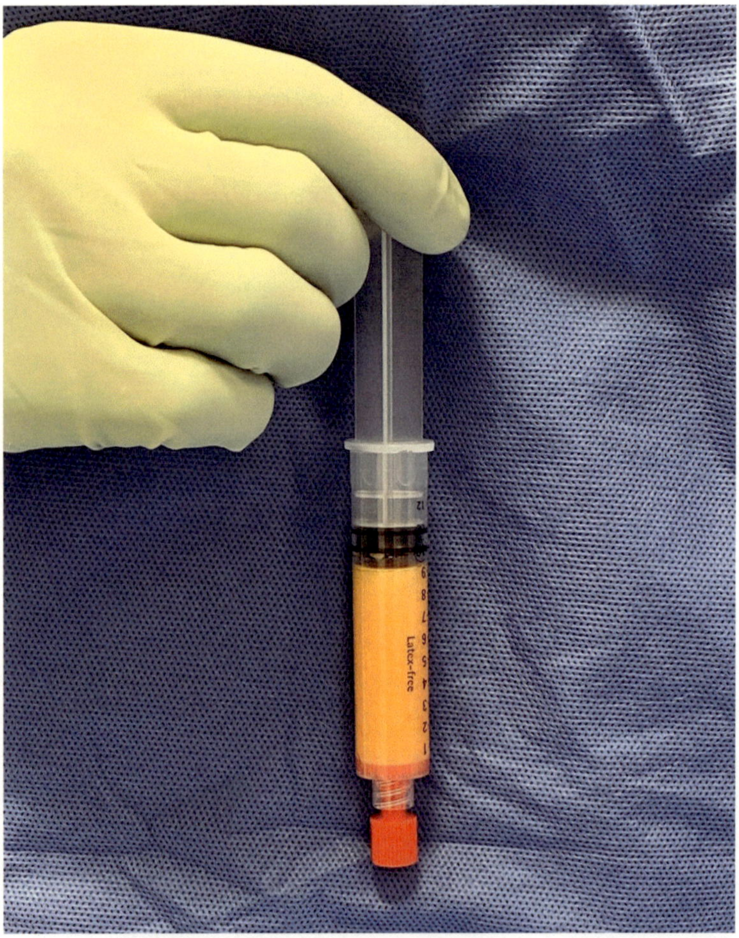

Figure 2.
Decantation and sedimentation allow the product to separate into three layers (aqueous, oil and fat). The oil and aqueous phases are then extracted, while the fat layer is withdrawn for injection.

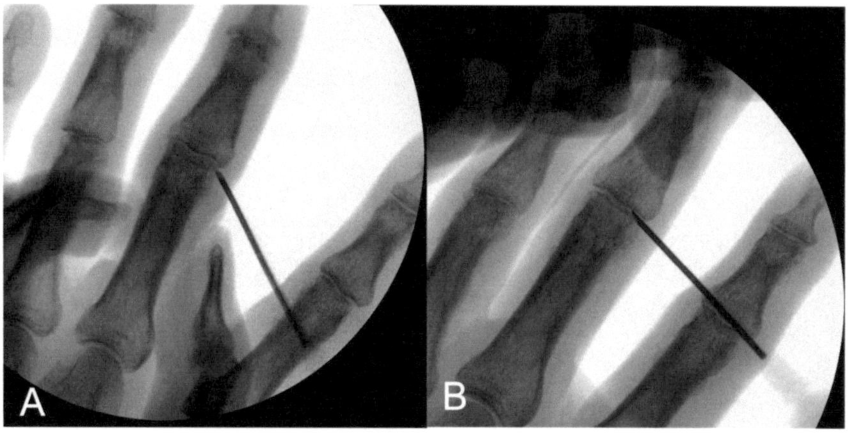

Figure 3.
Liparthroplasty of the proximal interphalangeal joint of the left middle finger in a 60-year-old man. A) Fluoroscopy before the injection. B) with an 18G needle and a 3 mL Luer-lock syringe 0.5 mL of lipoaspirate was injected into the joint. A widening of the joint is clearly visible after the injection.

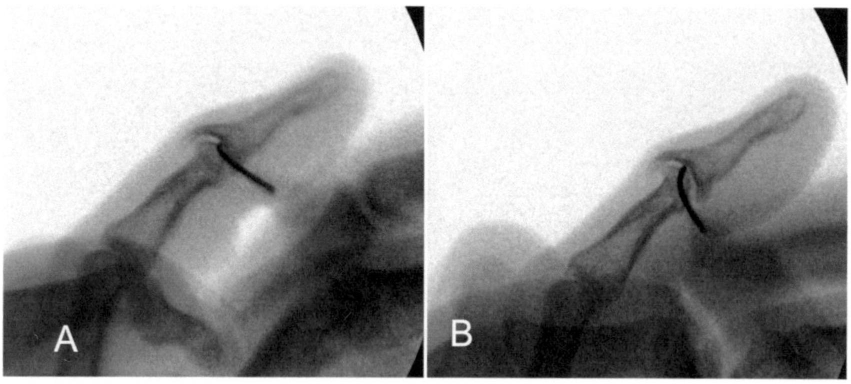

Figure 4.
Autologous fat infiltration of a distal interphalangeal joint of the right ring finger in a 63-year-old man. A) before infiltration. B) Fluoroscopy during injection of 0.5 mL of fat. A widening of the joint is clearly visible after the injection.

2.6 Fat processing technique

2.6.1 Centrifugation

Coleman [16] introduced a fat-processing technique for reinjection. This method, still widely used, involves the separation of fat components using high-speed centrifugation. Numerous research studies have documented varying centrifugation configurations [17]. Some of these studies propose that higher levels of centrifugal force could harm the fat cells, resulting in reduced cell viability. Conversely, very low centrifugal forces yield outcomes similar to the simple process of sedimentation [17]. Pulsfort et al. [18] indicated that variations in centrifugation techniques did not lead to significant histological changes in the viability of adipocytes and did not result in apoptotic changes.

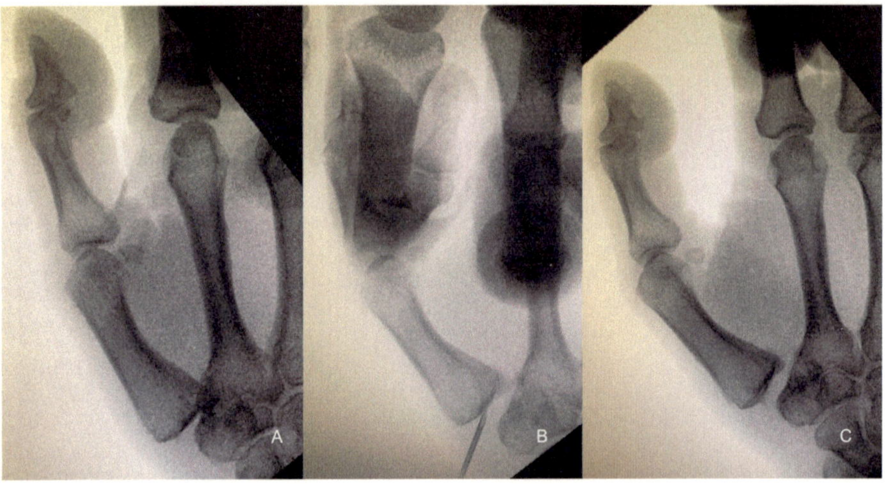

Figure 5.
Liparthroplasty of the right trapeziometacarpal joint in a 56-year-old woman with a painful Eaton stage I basal thumb osteoarthritis. A) Fluoroscopy before the injection. B) with an 18G needle and a 3 mL Luer-lock syringe 1.5 mL of lipoaspirate was injected into the trapezio-metacarpal joint. C) Fluoroscopy after the procedure without traction.

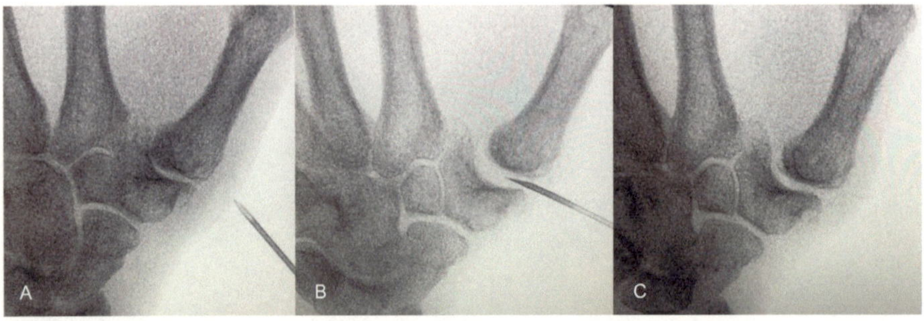

Figure 6.
Autologous fat infiltration of a left trapeziometacarpal joint of a 61-year-old male patient with an Eaton stage II basal thumb osteoarthritis. A) before infiltration. B) Fluoroscopy during injection of 1 mL of fat and C) at the end of the procedure.

2.6.2 Cotton gauze technique

The method of cotton gauze rolling is another commonly utilized approach to isolate harvested fat grafts. In this process, the gathered fat is positioned over the gauze. A forceps or tongue depressor is used to isolate the fat across the gauze. The gauze absorbs fluid and oil, leaving behind the cellular components of the fat graft. As the blood and other elements of lipoaspirate are removed, the harvested fat takes on a more pronounced yellow color. This technique is known to take approximately 2 to 4 minutes, it is cheap with minimal damage to the fat [19].

2.6.3 Decantation

Decantation and sedimentation allow the product to separate into three layers (aqueous, oil and fat). The oil and aqueous phases are then extracted, while the fat layer is withdrawn for injection [20].

2.6.4 Washing and filtration

Lipoaspirate preparation can involve washing and/or filtration, usually conducted within a closed system. These two methods can be carried out individually or in combination. Washing generally involves multiple cycles with lactated Ringer's solution, while filtration takes place through membranes of various pore sizes [21].

Currently, no single technique exhibits superior results in terms of fat graft take compared to others. Decantation leads to a higher number of viable adipocytes along with undesired cellular components, which ultimately results in lower graft take compared to centrifugation and washing methods [22].

3. Results

A total of 148 studies were identified; 17 additional articles were found through previous publications (a total of 165 articles). A total of 150 records resulted after duplicates were removed. A total of 14 studies were selected and only 10 respected the inclusion criteria (**Figure** 7).

3.1 Study characteristics

Of the 10 included studies, 9 reported on trapeziometacarpal joint injection. Only one was reported on osteoarthritic finger joints. **Table 1** summarizes the results.

The article included in the systematic review was case-control studies (n = 2), case series (n = 7) and a case report (n = 1). The harvested fat processing system was done by centrifugation (n = 6), mechanical homogenization alone (n = 1), mechanical homogenization and filtration (n = 1), decantation and mechanical homogenization (n = 2). None of the above-mentioned studies examined the injected tissue for cell count or viability. Follow-up ranged from 3 months to five years. Only one study analyzed metacarpophalangeal infiltrations of the proximal interphalangeal and distal interphalangeal fingers. In all studies, intra-articular injections of 0.5 mL up to 2 mL of fat into the joints were performed under fluoroscopic guidance.

3.1.1 Indications

The indications where osteoarthritis with radiological findings showing ostearthritis and failed medical treatment.

3.1.2 Pain

In terms of symptomatic pain relief (VAS) after fat grafting, outcomes have been consistently positive. Only one article [23] showed non-statistically significant improvement of the pain.

3.1.3 Strength

Five studies reported non-statistically significant improvement of the strength after the liparthroplasty [23, 26–29]. In 4 more recent studies [25, 30–32] with a larger court of patients the authors report statistically significant improvement in strength as well. There is a trend to an improvement of grip and pinch strength after 12 h months (**Table 1**).

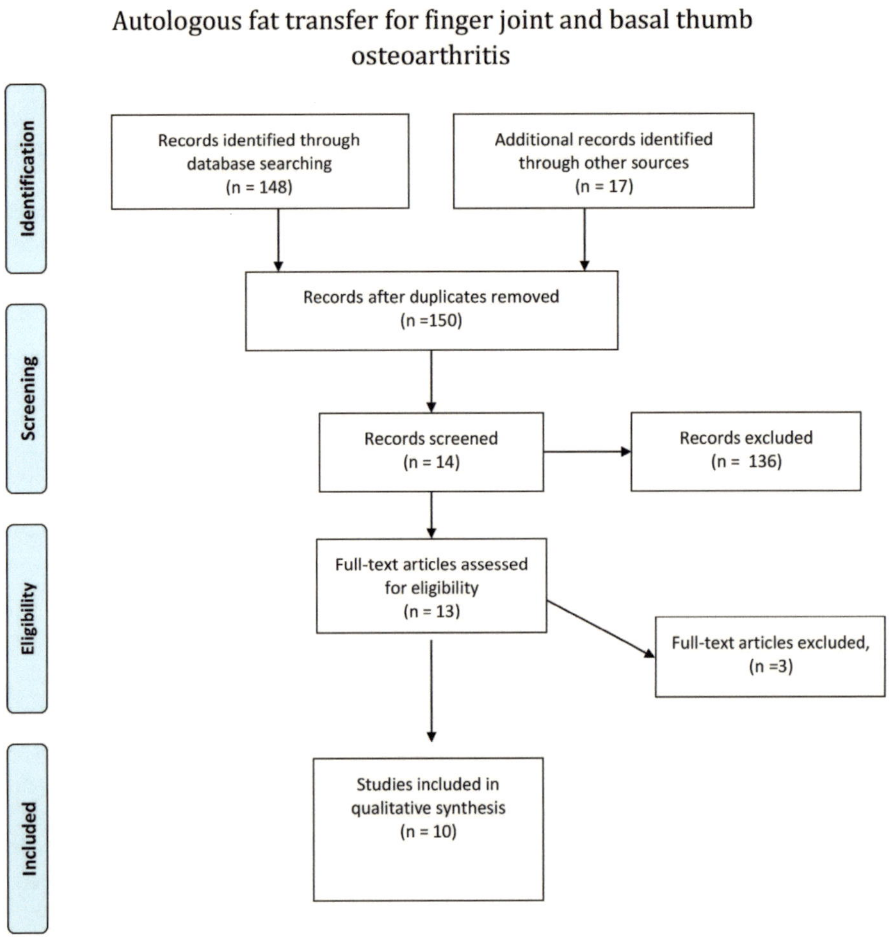

Figure 7.
PRISMA statement of the study.

3.1.4 Function of the hand

Statistically significant DASH Score (Disabilities of Arm, Shoulder and Hand) improvement after fat injection is reported in all the studies except one [27].

3.1.5 Complications

No significant complications were reported. One patient who had a hematoma in the donor site and three patients reported hypoesthesia in the skin area supplied by the superficial radial nerve.

3.1.6 Study analysis

Herold et al. [23] conducted a small case series involving five patients with thumb carpometacarpal joint (CMCJ) osteoarthritis (stages II and III). Three months after autologous fat injection, they observed improvements in grip strength, pinch strength and hand function as measured by the DASH score. However, these improvements

Author	Patients	Joint	Eaton classification	Fat grafting technique	Follow up	VAS	Pinch/Grip	DASH/MHQ	Weaknesses
Herold [23] 2014	5 patients	TM joint	II -III	Centrifugation (Coleman)	3 months	Non statistically significant reduction	Non statistically significant improvement	Significant improvement after liparthroplasty	Short follow up, lack of controls, small sample size
Bohr [24] 2015	1 patient	TM joint	II	Centrifugation (Coleman)	12 months	N/A	N/A	Significant improvement after liparthroplasty	Lack of clinical parameters, small sample size
Herold [25] 2017	50 patients	TM joint	II-IV	Centrifugation (Coleman)	12 months	Significant reduction after liparthroplasty	Significant improvement after liparthroplasty	Significant improvement after liparthroplasty	Lack of clinical parameters, small sample size
Haas [26] 2017	24 patients, 12 triamcinolone, 12 fat grafting	TM joint	I-III	Mechanical homogenization	3 months	Significant reduction after liparthroplasty	Non statistically significant improvement	Significant improvement after liparthroplasty	Short follow up, Lack of controls
Erne [27] 2018	21 patients, 9 fat grafting, 12 resection arthroplasty	TM joint	III-IV	Centrifugation (Coleman)	12 months	Improvement of VAS but no significant difference	Non statistically significant improvement	Non statistically significant improvement	No randomization, small sample size
Haas [28] 2019	99 patients	TM joint	I-III	Filtration and mechanical homogenization	12 months Significant reduction after liparthro-plasty	Non statistically significant improvement	Significant improvement after liparthroplasty	Lack of controls	
Forschauer [29] 2020	31 patients	TM joint	II-III	Decantation/mechanical homogenzation	2 years	Significant reduction after liparthroplasty	Non statistically significant improvement	Significant improvement after liparthroplasty	Lack of controls

Author	Patients	Joint	Eaton classification	Fat grafting technique	Follow up	VAS	Pinch/Grip	DASH/MHQ	Weaknesses
Herold [30] 2022	42 patients	TM joint	II -III	Centrifugation (Coleman)	4 years	Significant reduction after liparthroplasty	Significant improvement after liparthroplasty	Significant improvement after liparthroplasty	Lack of controls
Meyer-Marcotty [31] 2022	25 patients	MP joint n = 4 PIP joint n = 18 DIP joint n = 3	N/A	Centrifugation (Coleman)	44 months (3.7 years)	Significant reduction after liparthroplasty	Significant improvement of Pinch after liparthroplasty	Significant improvement after liparthroplasty	Lack of controls
Holzbauer [32] 2022	31 patients	TM joint	II-III	Decantation/mechanical homogenzatio	5.1 years	Significant reduction after liparthroplasty	Significant improvement of Pinch after liparthroplasty	Significant improvement after liparthroplasty	Lack of controls

Table 1.
Comparison of all the article from the literature for the liparthroplasty of trapeziometacarpal joint.

were not statistically significant compared with preoperative values. The study had limitations, such as short follow-up and a limited sample size. The same authors reported on greater case series with 50 patients and a follow-up of one year, reporting significant reductions in pain, improvements in grip and pinch strength and Quick-DASH scores after intra-articular fat injection.

Bohr et al. [24] reported improvement in DASH scores in a single case of stage II CMCJ arthritis treated with intra-articular fat injection. However, this study is limited by being a single case report without objective clinical evaluation and control groups.

In the first case-control study conducted in 2017 by Haas et al. [26], patients with stage I-III CMCJ arthritis who received autologous fat transplantation or corticosteroid injection were examined. The fat processing technique involved mechanical homogenization, but no details were given about the specific fat product used. After 3 months, the fat group experienced a significant reduction in pain and improvement in quality of life (Quick DASH and Michigan Hand Questionnarire-MHQ scores), while the corticosteroid group, although initially improved, regressed below preoperative levels after six weeks. Grip strength remained unchanged in both groups, and the follow-up period was relatively short.

Erne et al. [27] conducted a case-control study of 21 patients with advanced CMCJ arthritis of the thumb (Eaton stages III and IV). They compared autologous fat grafting with Lundborg resection arthroplasty, with patients given a choice between the less invasive and traditional surgical approaches. The group of patients treated with fat experienced a significantly shorter time to pain resolution (1.7 months compared with 5.7 months in the group of patients treated with resection). Pain levels (Visual Analogue Scale - VAS), grip and pinch strength and DASH scores improved similarly in both groups after 12 months. The duration of the intervention was also significantly shorter in the fat group. However, this study had limitations, such as a small sample size and lack of randomization, which introduced a potential selection bias.

Herold et al. in 2017 reported on a wider group involving 50 patients with one-year-follow-up [25]. In this study, intra-articular lipofilling resulted in a significant reduction in pain, higher values of grip and pinch strength and a lower Quick DASH score.

In 2019 Haas et al. [28] published a study with the largest number of patients, involving 99 patients with symptomatic CMCJ osteoarthritis of the thumb. These patients received 1–2 mL of mechanically homogenized autologous fat. After 12 months, the study demonstrated significant pain reduction and improvements in DASH and MHQ scores. However, grip and pinch strength remained unchanged.

Forschauer et al. [29] in 2020 conducted a two-year follow-up study of 31 patients with stage II and III osteoarthritis. They reported results consistent with other studies, including a significant reduction in pain and improvement in grip and pinch strength following autologous fat injection.

In 2022 Herold et al. reported on a group of 42 patients with a longer follow-up, averaging 4.9 ± 0.7 years [30]. The authors noted that, among the initial 50 consecutive patients, three needed surgical conversion within the first year after the intervention. In comparison to the previously mentioned study [25] they present slightly elevated pain levels during activity (4.0 ± 3.0) and DASH Scores (31 ± 22), although these outcomes remained significantly lower than the pre-intervention levels.

In 2022 Meyer-Marcotty [31] investigated the autologous fat injection of 25 finger joints of the hand with a mean follow-up of 44 months. The median DASH score improved non-significantly from 50 (3 t ± 72) to 25 (0 ± 85). The median level of pain experienced showed a highly significant improvement from 6.0 (1.0 ± 10.0) to 0.5 (0.0 ± 6.5). The median force of pinch grip rose highly significantly from 2.00 kg (0.00 ± 11.00 kg) to 4.30 kg (2.00 ± 12.00 kg).

Holzbauer et al. [32] described 31 patients treated with fat infiltration of the TM joint after a follow-up period of 5.1 years. They reported a tendency of surgical conversion after the lipofilling, so that 61% of 31 initially infiltrated patients could be studied in this chapter. An important result of this study is that smoking have a significantly higher risk for therapy failure.

4. Discussion

Since the inception of autologous lipoaspirated fat transplantation as a therapy for osteoarthritis (OA) of the trapeziometacarpal joint [23] numerous research groups have explored this innovative approach involving the injection of autologous fat. Zuk et al. [33] were the first to describe the presence of mesenchymal stem cells in adipose tissue in 2001. The mesenchymal stem cells have regenerative abilities and a potential to differentiate into multiple cell types, which is comparable to bone marrow mesenchymal stem cells [1–4]. Adipose-derived SVF obtained through enzymatic digestion and centrifugation of collected fat tissue has gained significant popularity for clinical applications in joints displaying osteoarthritic changes [6–9]. This SVF comprises a diverse cellular composition, including mesenchymal stem cells, endothelial progenitor cells, pericytes and immune cells. It shows strong regenerative potential and the possibility of synergistic impact on immunomodulation, inflammation, and angiogenesis. Adipose-derived stem cells (ASCs) are typically extracted from the stromal vascular fraction (SVF) of adipose tissue. Their morphology is similar to fibroblasts and shows positivity for common mesenchymal stem cell (MSC) markers such as CD44, CD73, CD90 and CD105, while lacking expression of typical hematopoietic markers such as CD14, CD34 and CD45 [10, 12, 33]. Like MSCs, ASCs have the ability to differentiate into various lineages, including osteogenic, adipogenic, myogenic and chondrogenic [12]. Regenerative fat cells exert their regenerative effects primarily through paracrine action [34]. They secrete bioactive molecules, such as cytokines and growth factors, which affect neighboring cells and tissues. These molecules can promote angiogenesis, reduce inflammation and modulate the immune system [12–14]. Adipose-derived exosomes, which are small vesicles secreted by adipose stem cells, contain various bioactive molecules, including proteins and microRNAs. Studies have shown that adipose-derived exosomes have anti-inflammatory and tissue regenerative effects [35].

The examined studies on intra-articular autologous fat injection for basal thumb and finger osteoarthritis are somewhat limited, but their results are undeniably positive with a great potential for treatment for many years. The injection of autologous fat into the joint consistently led to a reduction in pain and subsequent improvements in hand function. However, when it comes to grip and pinch strength, the outcomes were not consistent but there is a trend to an improvement 12 months after treatment.

Intra-articular injections of cortisone or hyaluronic acid serve as alternative approaches to the liparthroplasty. It's important not to underestimate the potential chondral damage resulting from repeated cortisone injections [3, 5]. With the liparthroplasty, several other surgical treatments remain available. Autologous fat transfer seems to be a secure and minimally invasive technique that does not burn any bridge to other techniques.

5. Conclusion

Fat grafting and adipose-derived cellular therapies can be effective in finger and thumb osteoarthritis treatment due to their biological and immunomodulatory characteristics. Although basic research provides evidence of these proprieties of adipose-derived stem cells [9–14], the exact mode of action of liparthroplasty is still not clear. Adipose tissue can provide a cushioning effect due to its ability to absorb and distribute mechanical forces. Based on the findings of this analysis, autologous fat injections into the affected finger joint seem to alleviate pain and enhance hand function. Further investigations should concentrate on identifying the most effective method for processing fat, enhancing the biological qualities and effectiveness of the injected cells. It will be imperative to conduct randomized controlled trials to ascertain whether the injection of autologous fat grafting could significantly contribute to the symptomatic treatment of thumb and finger osteoarthritis.

Acknowledgements

This research received no external funding.

Conflict of interest

The authors declare no conflict of interest.

Author details

Marco Guidi[1*], Lucian Lior Marcovici[2], Jaime Sánchez Ruas[3], Camilo Chaves[4,5], Fabio Andrés Tandioy-Delgado[6], Ahlam Arnaout[7], Carlos Federico Morales Hackett[8], Stefano Lucchina[9], Enrico Carità[10] and Ivan Tami[1]

1 Hand Surgery Center Manoegomito, Clinica Ars Medica, Gravesano, Switzerland

2 Hand and Microsurgery Unit, Jewish Hospital of Rome, Rome, Italy

3 Dr. Piñal and Associates, Calle Serrano, Madrid, Spain

4 Clinique du Pré, Le Mans, France

5 Clinique Ambroise Paré, Neuilly-sur-Seine, France

6 Hand Surgery, Clinica las Américas, Medellin, Antioquia, Colombia

7 Bizet Clinic, Hand Institute, Paris, France

8 Hospital San Juan de Dios, Curico, Chile

9 Hospital La Carita' – Locarno Hand Center, Locarno, Switzerland

10 Hand and Microsurgery Unit, Clinica San Francesco, Verona, Italy

*Address all correspondence to: marcoguidi984@hotmail.com

References

[1] Bühler M, Chapple CM, Stebbings S, Pōtiki-Bryant K, David BG. Impact of thumb carpometacarpal joint osteoarthritis: A pragmatic qualitative study. Arthritis Care Research (Hoboken). 2021;**73**(3):336-346

[2] Riley N, Vella-Baldacchino M, Thurley N, Hopewell S, Carr AJ, Dean BJF. Injection therapy for base of thumb osteoarthritis: A systematic review and meta-analysis. BMJ Open. 2019;**9**:e027507

[3] Wernecke C, Braun HJ, Dragoo JL. The effect of intra-articular corticosteroids on articular cartilage: A systematic review. Orthopedic Journal of Sports Medicine. 2015;**3**(5)

[4] Suntiparpluacha M, Tammachote N, Tammachote R. Triamcinolone acetonide reduces viability, induces oxidative stress, and alters gene expressions of human chondrocytes. European Review for Medical and Pharmacological Sciences. 2016;**20**:4985-4992

[5] Zhang Y, Chen X, Tong Y, Luo J, Bi Q. Development and Prospect of intra-articular injection in the treatment of osteoarthritis: A review. Journal of Pain Research. 2020;**13**:1941-1955

[6] Qureshi MK, Halim UA, Khaled AS, Roche SJ, Arshad MS. Trapeziectomy with ligament reconstruction and tendon interposition versus trapeziometacarpal joint replacement for thumb carpometacarpal osteoarthritis: A systematic review and meta-analysis. Journal of Wrist Surgery. 2022;**11**:272-278

[7] Holme TJ, Karbowiak M, Clements J, Sharma R, Craik J, Ellahee N. Thumb CMCJ prosthetic total joint replacement: A systematic review. EFORT Open Review. 2021;**6**:316-330

[8] Raj S, Clay R, Ramji S, Shaunak R, Dadrewalla A, Sinha V, et al. Trapeziectomy versus joint replacement for first carpometacarpal (CMC 1) joint osteoarthritis: A systematic review and meta-analysis. European Journal of Orthopaedic Surgery and Traumatology. 2022;**32**(6):1001-1021

[9] Jo CH, Lee YG, Shin WH, Kim H, Chai JW, Jeong EC, et al. Intra-articular injection of mesenchymal stem cells for the treatment of osteoarthritis of the knee. A proof of concept clinical trial. Stem Cells. 2014;**32**:1254-1266

[10] Bora P, Majumdar AS. Adipose tissue-derived stromal vascular fraction in regenerative medicine: A brief review on biology and translation. Stem Cell Research & Therapy. 2017;**8**:145

[11] ter Huurne M, Schelbergen R, Blattes R, Blom A, de Munter W, Grevers LC, et al. Antiinflammatory and chondroprotective effects of intraarticular injection of adipose-derived stem cells in experimental osteoarthritis. Arthritis and Rheumatism. 2012;**64**:3604-3613

[12] Gimble JM, Katz AJ, Bunnell BA. Adipose-derived stem cells for regenerative medicine. Circulation Research. 2007;**100**:1249-1260

[13] Michalek J, Moster R, Lukac L, Proefrock K, Petrasovic M, Rybar J, et al. Stromal vascular fraction cells of adipose and connective tissue in people with osteoarthritis: A case control prospective multi-centric non-randomized study. Global Surgery. 2017;**3**:1-9

[14] Feisst V, Meidinger S, Locke MB. From bench to bedside: Use of human adipose-derived stem cells. Stem Cells Cloning. 2015;**8**:149-162

[15] Liberati A, Altman DG, Tetzlaff J, Mulrow C, Gøtzsche PC, Ioannidis JP, et al. The PRISMA statement for reporting systematic reviews and meta-analyses of studies that evaluate healthcare interventions: Explanation and elaboration. BMJ. 2009;**339**:b2700

[16] Coleman SR. Structural fat grafts. Clinics in Plastic Surgery. 2001;**28**:111-119

[17] Strong AL, Cederna PS, Rubin JP, Coleman SR, Levi B. The current state of fat grafting: A review of harvesting, processing, and injection techniques. Plastic and Reconstructive Surgery. 2015;**136**(4):897-912

[18] Pulsfort AK, Wolter TP, Pallua N. The effect of centrifugal forces on viability of adipocytes in centrifuged lipoaspirates. Annals of Plastic Surgery. 2011;**66**(3):292-295

[19] Fisher C, Grahovac TL, Schafer ME, Shippert RD, Marra KG, Rubin JP. Comparison of harvest and processing techniques for fat grafting and adipose stem cell isolation. Plastic and Reconstructive Surgery. 2013;**132**:351-361

[20] Rose JG, Lucarelli MJ, Lemke BN, et al. Histologic comparison of autologous fat processing methods. Ophthalmic Plastic and Reconstructive Surgery. 2006;**22**:195-200

[21] Zhu M, Zhou Z, Chen Y, et al. Supplementation of fat grafts with adipose-derived regenerative cells improves long-term graft retention. Annals of Plastic Surgery. 2010;**64**:222-228

[22] Pfaff M, Wu W, Zellner E, Steinbacher DM. Processing technique for lipofilling influences adipose-derived stem cell concentration and cell viability in lipoaspirate. Aesthetic Plastic Surgery. 2014;**38**:224-229

[23] Herold C, Fleischer O, Allert S. Eigenfettinjektion in das Sattelgelenk zur Behandlung der Rhizarthrose—eine viel ver- sprechende Therapieoption. Handchirurgie, Mikrochirurgie, Plastische Chirurgie. 2014;**46**:108-112

[24] Bohr S, Rennekampff HO, Pallua N. Cell-enriched Lipoaspirate arthroplasty: A novel approach to first carpometacarpal joint arthritis. Hand Surgery. 2015;**20**:479-481

[25] Herold C, Rennekampff HO, Groddeck R, Allert S. Autologous fat transfer for thumb carpometacarpal joint osteoarthritis: A prospective study. Plastic and Reconstructive Surgery. 2017;**140**:327-335

[26] Haas EM, Volkmer E, Giunta RE. Pilotstudie über die Wirkung und den Nutzen von autologen Fettgewebstransplantaten bei Rhizarthrose verglichen mit einer Kortisoninjektion—3 Monatsergebnisse. Handchirurgie, Mikrochirurgie, Plastische Chirurgie. 2017;**49**:288-296

[27] Erne HC, Cerny MK, Ehrl D, Bauer AT, Schmauss V, Moog P, et al. Autologous fat injection versus Lundborg resection arthroplasty for the treatment of Trapeziometacarpal joint osteoarthritis. Plastic and Reconstructive Surgery. 2018;**141**:119-124

[28] Haas EM, Eisele A, Arnoldi A, Paolini M, Ehrl D, Volkmer E, et al. One-year outcomes of Intraarticular fat transplantation for thumb carpometacarpal joint osteoarthritis: Case review of 99 joints. Plastic and Reconstructive Surgery. 2020;**145**:151-159

[29] Froschauer SM, Holzbauer M, Wenny R, Schmidt M, Huemer GM, Kwasny O, et al. Autologous fat transplantation for thumb carpometacarpal joint osteoarthritis

(Liparthroplasty): A case series with two years of follow-UP. Journal of Clinical Medicine. 2020;**10**:113

[30] Herold C, Lange J, Rennekampff HO, Allert S, Meyer Marcotty M. Eigenfettinjektion bei Rhizarthrose: 5 Jahresergebnisse bei 42 Patienten. Z. Orthop. Unfall. 2022

[31] Meyer-Marcotty M, Batsilas I, Sanders A, Dahmann S, Happe C, Herold C. Lipofilling in osteoarthritis of the finger joints: Initial prospective long-term results. Plastic and Reconstructive Surgery. 2022;**149**(5):1139-1145

[32] Holzbauer M, Hopfner M, Haslhofer D, Kwasny O, Duscher D, Froschauer SM. Radial and palmar active range of motion measurement: Reliability of six methods in healthy adults. Journal of Plastic Surgery and Hand Surgery. 2021;**55**:41-47

[33] Zuk PA, Zhu M, Ashjian P, De Ugarte DA, Huang JI, Mizuno H, et al. Human adipose tissue is a source of multipotent stem cells. Molecular Biology of the Cell. 2002;**13**(12):4279-4295

[34] Trzyna A, Banaś-Ząbczyk A. Adipose-derived stem cells Secretome and its potential application in "stem cell-free therapy". Biomolecules. 2021;**11**(6):878

[35] Heo JS, Kim S, Yang CE, Choi Y, Song SY, Kim HO. Human adipose mesenchymal stem cell-derived exosomes: A key player in wound healing. Tissue Engineering and Regenerative Medicine. 2021;**18**(4):537-548